Handbook of Food Science and Technology 2

Series Editor
Jack Legrand & Gilles Trystram

Handbook of Food Science and Technology 2

Food Process Engineering and Packaging

Edited by

Romain Jeantet
Thomas Croguennec
Pierre Schuck
Gérard Brulé

iSTE

WILEY

First published 2016 in Great Britain and the United States by ISTE Ltd and John Wiley & Sons, Inc.
Translated by Geraldine Brodkorb from "Science des aliments" © Tec & Doc Lavoisier 2006.

ISTE Ltd
27-37 St George's Road
London SW19 4EU
UK

www.iste.co.uk

John Wiley & Sons, Inc.
111 River Street
Hoboken, NJ 07030
USA

www.wiley.com

Library of Congress Control Number: 2016930387

British Library Cataloguing-in-Publication Data
A CIP record for this book is available from the British Library
ISBN 978-1-84821-933-5

Contents

Introduction

Food is a complex and heterogeneous system. It often consists of a protein and/or polysaccharide matrix surrounding, sustaining a typically aqueous continuous phase containing soluble hydrophilic compounds (carbohydrates, salts, vitamins, etc.) and some dispersed elements (cells, fat globules, gas bubbles, crystals, etc.) (Figure I.1). Such a system is thermodynamically, biologically and chemically very unstable.

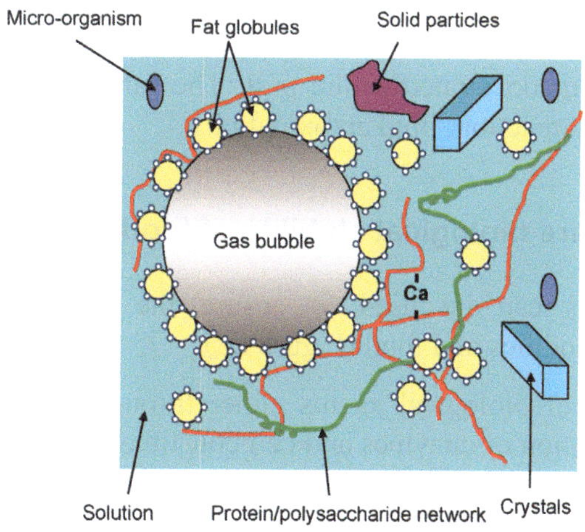

Figure I.1. *Matrix structure of food*

Introduction written by Gérard BRULÉ.

The dispersed elements are subject to forces that cause phase separations either by sedimentation when the density of dispersed elements is greater than that of the dispersing phase, or by creaming when the opposite applies. Chemical potential or pressure gradients exerted on either side of the interfaces can induce the transfer of solutes and structural changes (coalescence, plasmolysis). This physicochemical instability can be increased by mechanical and thermal stresses on products during storage (refrigeration, freezing) and preparation (defrosting, reheating, etc.).

Protein and polysaccharide polymers that contribute significantly to the structuring of food are likely to reorganise during storage because of the influence of temperature on hydrophobic, ionic and hydrogen interactions. As described in Volume 1 [JEA 16a], recrystallization is sometimes accompanied by the release and migration of water with textural changes (starch retrogradation and bread staling).

Biological agents, enzymes and microorganisms find suitable conditions for their action and development in most foods both in terms of physico-chemical conditions (pH, water activity a_w and temperature) as well as availability of substrates and growth factors. Lipolysis, proteolysis and oxidation reactions, metabolite production (acid, alcohol, gas), the development of pathogenic flora and the release of toxins are all elements that negatively affect the sensory, safety and nutritional quality of food, as has already been demonstrated in the past. The same applies for chemical reactions (Maillard reaction, lipid autoxidation).

I.1. How to ensure biological stability of foods?

Four strategies can be used to ensure the biological stability of agricultural commodities and food (Figure I.2):

– inactivation of biological agents either through energy input, which induces a denaturation of enzymes and cell constituents, or through physical, chemical or enzymatic treatments (latter being able to lyse or alter some transport properties of the cell wall);

– separation of biological agents based on density differences or their size;

– inhibition of enzymes, microorganisms and reactions by decreasing water availability (lowering a_w), which limits both the transfer of substrates and growth factors as well as the transfer of metabolites or reaction products that accumulate in the reaction environment;

– Inhibition through the creation of limiting physicochemical conditions (pH, oxygen pressure, bacterial inhibitors, etc.).

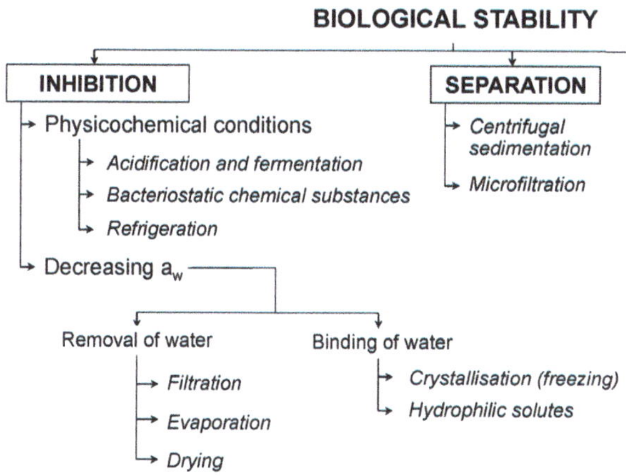

Figure I.2. *Strategies for the biological stability of food*

All these strategies will be detailed in Part 2 of this book.

I.1.1. *Thermal inactivation of biological agents*

The destruction of microorganisms and the inactivation of enzymes by heat treatments that provide energy for the denaturation of macromolecules (proteins, DNA) were first applied by Appert and Pasteur. Food pasteurization and sterilization have been defined taking into account the thermal sensitivity of pathogenic flora:

– pasteurization destroys a large proportion of undesirable flora (*Listeria*, *Staphylococci*, *Salmonella sp.*, etc.) and extends shelf life in the cold chain;

– sterilization destroys all microorganisms and denatures almost all enzymes (depending on the heat treatment) to allow long-term storage at room temperature.

To reduce the risk of pathogen proliferation, treatments are based on the thermal resistance of *Clostridium botulinum*, one of the most heat stable and pathogenic microorganisms. By aiming to destroy microorganisms and inactivate enzymes through thermal denaturation, macromolecules responsible for food texture and biological constituents (vitamins) are also partially or totally denatured. These side effects degrade the sensory and nutritional quality of food. Thus, successfully destroying microorganisms without reducing the food quality too much requires the precise control of heat transfers and momentum during heat treatments. Heat treatment is not always used due to the thermal sensitivity of some products (e.g. egg white) or in order to preserve the raw nature of the product. Other non-thermal treatments (i.e. inducing a temperature increase of less than 25°C) are therefore used, for example ionization, high pressure and pulsed electric fields. These techniques have limited applications mainly due to cost, legislation and consumer acceptability. Hurdle technologies (combination of treatments to minimise energy input in the food product) may be an alternative.

I.1.2. *Inhibition by decreasing water activity*

Most microbial and biochemical changes that reduce the quality of food occur in the aqueous phase. Water has a dual role:

– as a solvent, it ensures the transfer of substrates, growth factors, biological agents and reaction products thereby creating optimum conditions for reactions;

– as a reaction substrate, it is involved in hydrolysis reactions (proteolysis, lipolysis).

This dual action of water requires its availability, which is characterised by a_w: any treatment that reduces water availability slows down the reaction rate. a_w can be lowered by the removal of free water through evaporation or drying, the crystallization of solvent water (freezing) or the addition of highly hydrophilic solutes that bind water molecules through hydrogen or dipolar interactions (salting, sugaring). Inhibition through the decrease of a_w can be reversed by rehydration, defrosting and dilution. This method of preservation only slightly affects the nutritional and sensory quality if freezing and dehydration are well controlled in terms of heat and mass transfers. The shelf life may however be limited in the case of fat-containing products since reactions involving lipid substrates occur at the interface between the lipid and aqueous phases, and are consequently less dependent on water availability. Conversely, lipid oxidation is increased when a_w reaches 0, as exposure to oxygen is increased in these conditions.

I.1.3. Chemical inhibition

The kinetics of microbial growth and the rate constants of enzymatic reactions are highly dependent on the pH of the medium. It is possible to slow down biological phenomena by deviating from the optimum reaction pH. As this is around neutral pH for most microorganisms and enzymes, microbial growth can be limited by acidifying the medium either through the addition of acid (acetic acid, lactic acid, propionic acid, carbonic acid, etc.) or through fermentation. Aerobic microorganism growth can also be controlled by reducing the oxygen pressure through vacuum packaging or modified atmosphere packaging.

Many metabolites and reaction products are themselves inhibitors and, above a certain concentration, can slow down or terminate fermentation (alcohol, lactic acid or propionic acid) or reactions.

The bacteriostatic or bactericidal role of some chemical substances has been shown in ancestral processes: this is the case with nitrate/nitrite in meat curing and phenolic compounds resulting from the thermal degradation of lignin in the smoking process. These technologies generate undesirable substances due to their carcinogenic nature (nitrosamines, benzopyrenes). A good understanding of the biochemical reactions and processing conditions responsible for the formation of these substances can limit their levels in food.

I.2. How to ensure physicochemical stability of foods

The presence of dispersed elements in food (fat globules in the case of emulsions, air bubbles in the case of foams) or the presence of proteins and carbohydrates, which are thermodynamically incompatible, can destabilise the product. The chemical potential gradients at interfaces can induce water and solute transfers between two phases or between a base and a topping or filling (e.g. hydration of a shortbread or puff-pastry base from the topping/filling); all these transfers are able to reduce the sensory quality of food.

The transfer rate v of dispersed elements is governed by Stockes law:

$$\mathrm{v} = \frac{D^2}{18\eta}\Delta\rho \mathrm{g} \qquad [I.1]$$

with D the diameter of particles (m), $\Delta\rho$ the density difference between the dispersed phase and the continuous phase (kg m^{-3}), η the dynamic viscosity of the continuous phase (Pa s) and g the gravitational acceleration (m s^{-2}).

To stabilise the dispersed elements, the parameters D and η can be adjusted, or a macromolecular network can be created to limit the movement of particles by entrapping them within the network.

The diameter of dispersed elements like fat globules can be reduced by mechanical treatments such as homogenization: in order to obtain relatively small diameters, it is necessary to lower the interfacial tension that creates resistance to an increase in the interphase surface. This surface tension can be lowered using amphiphilic molecules positioned at the interface, the hydrophobic portion facing the lipid phase and the hydrophilic portion facing the aqueous phase. In addition to creating micro-emulsions, it is also essential that the fat globules do not agglomerate, which could result in coalescence and an increase of the fat globule diameter. To avoid this, ionic and steric constraints must be created on the surface of the globule, which by repulsion limit the coalescence of particles. Reducing D is not always enough to stabilise the system; it is also possible to increase viscosity by increasing the concentration of solutes (e.g. adding sugars) or by immobilising solvent water by adding carbohydrate or protein macromolecules to induce thickening.

It is more difficult to stabilise a foam than an emulsion since the density difference between the dispersed and the dispersing phases is 1,000 kg m^{-3} in a foam whereas it is 100 kg m^{-3} in an emulsion. It is therefore necessary to create a macromolecular network that immobilises air bubbles. Globular proteins, especially from egg whites, have good foaming properties since they position themselves at the [air/water] interface during beating, which creates a dissymmetry in their environment. As a result, denaturation occurs through changes in their tertiary or even secondary structures, which may be accompanied by molecular interactions and help forming a cohesive film ensuring good foam stability.

The dynamics of water and solutes between the various components of food can be controlled by creating hydrophobic interfaces; this is the role of fat in puff or shortcrust pastry. It is also necessary that the physical properties of this fat are not adversely affected by the crystallization of triglycerides during the refrigeration of the products.

The physicochemical stability of a product must be controlled throughout its shelf life but also during its handling by consumers including defrosting and reheating. Controlling the stability of a complex system that is thermodynamically unstable is one of the major challenges facing food engineers; this issue will be addressed in Part 3 of this book. Nowadays, several functional carbohydrate and protein ingredients are available to limit transfers and destabilization within the food matrix. Part 4 of this book will be devoted to the physicochemical basis of fractionation and related technologies used in the development of functional ingredients.

I.3. Packaging, an essential attribute for food preservation

When responding to market needs and consumer demands with regards to safety, sensory and nutritional quality, offering a service and preserving, where possible, the traditional character of food is a constant challenge for the food industry. Quality control requires a good understanding of the food and its biochemical, chemical and thermodynamic reactivity as well as the processing and storage methods involved. Indeed, food can be transported over long distances, stored, distributed and finally endure a sufficiently long shelf life in the hands of the consumer. Hence, the need for packaging to protect the product is something that satisfies specific functions in terms of food preservation. Packaging could also be described as the last stage of the production chain. These elements will be addressed in Part 5 of this book.

PART 1

Basis of Food Engineering

1

Transport Phenomena – Basis of Unit Operations

The processing of agricultural commodities into finished food products can be analyzed in different ways. It can be examined by sector or by establishing a detailed description of the different unit operations in each one. Such an approach is not very useful in that the same unit operations exist in different sectors. For example, there is very little difference, with the exception of product characteristics or the type of material used, between pasteurization in a brewery and that in a milk factory, between concentration in a milk factory and that in a sugar refinery or between water extraction of sugar from sugar beet and solvent extraction of fat from oilcake.

Such overlaps can be limited by another approach, which looks at processing at the level of elementary operations (or unit operations) such as filtration, centrifugation, pasteurization or drying.

However, this approach can be further rationalized as such unit operations are fundamentally linked to three types of transport phenomena (or transfers) resulting from a physical difference (temperature, concentration or speed) between two points within a system: heat transfer (transfer of thermal energy from a hot to a cold point), mass transfer (transfer of mass from a concentrated to a diluted point) and momentum

---

Chapter written by Romain JEANTET.

transfer (transfer of momentum from a moving point at a given speed to a moving point at a lower speed).

For example:

– heat transfers play a major role in the elementary operations of pasteurization, sterilization and concentration;

– mass transfers mainly govern extraction operations by diffusion.

However, such transfer processes are often connected and interdependent, for example in a drying operation, heat transfer causes the mass transfer (of moisture) from a product to the environment.

It is therefore essential to first examine these before studying unit operations, which combined together form the basis for food production overall.

There are three different mechanisms of heat transfer:

– *conduction (or diffusion)*: in this case, the matrix in which heat transfer takes place, can be considered motionless with regard to the direction of transfer. For example, heat transfer in a solid does not involve motion of the material itself. In a liquid, heat transfer may be conductive if the energy is transferred by molecular agitation, apart from the random motion of molecules in the direction of the heat transfer;

– *convection*: in this case, transfer processes involve movement of part of a matrix that carries a quantity of heat, mass or momentum from one point to another. For example, central heating is convective on two levels: heat produced by the boiler is transported to radiators by the forced circulation of water in the pipes, and from the radiators to the room by the convective movement of air;

– *radiation*: this type of transfer, which is specific to heat transfers, occurs by electromagnetic wave propagation and does not require material support. For example, browning of foods can be carried out by infrared radiation.

These three transport mechanisms can coexist. Radiation may be considered dominant in the case of very high temperatures. However, in most cases in the food industry, the temperatures used (between –20°C and

+150°C) are relatively low and radiation can be disregarded. As such, conduction and convection processes will be the focus of attention in this chapter.

1.1. Transfer processes in conduction

In conduction, transfer rates are linked to the characteristics of the particular system and are directly proportional to the exchange surface area and the potential gradient.

As a result, there is a close similarity between the equations describing heat, mass and momentum transfers.

1.1.1. *Heat transfers: Fourier's law*

Fourier's law is expressed by the vector relationship:

$$\vec{\phi_q} = -\lambda\ \overrightarrow{\mathrm{grad}(\theta)} \qquad [1.1]$$

$\vec{\phi_q}$ is the heat flux density vector (W m^{-2}), i.e. the scalar value $\frac{1}{dA}\frac{dQ}{dt}$ (dA: elementary surface exchange; m^2) multiplied by the normal unit vector at a constant-temperature surface:

$$\phi_q = \frac{1}{dA}\frac{dQ}{dt} = -\lambda\frac{d\theta}{dx} \qquad [1.2]$$

The proportionality factor λ between the heat flux density and the temperature gradient is the thermal conductivity (W m^{-1} K^{-1}), characteristic of a given product.

1.1.1.1. *Establishing steady state conditions at an infinite plate*

Generally, in a system, physical quantities vary with time from one point to another. This system is in a non-steady state (or transient state). However, if the environment external to the system is invariant (constant), the system, after a certain time, tends towards a state of equilibrium, where at any given point the physical quantities (in this case temperature) become invariant (constant) over time. Such a system has reached a steady state (or a stationary state).

The geometric model most often used to describe such a system is the semi-infinite plate (small thickness relative to length and width) which is assumed to be a homogeneous and isotropic solid, bounded by two parallel planes. The model includes the wall through which thermal energy is exchanged via a plate heat exchanger, which can be considered infinite as the edges have a negligible effect on heat transfer.

In this model, constant-temperature surfaces are parallel to the sides of the plate, and equation [1.2] is integrated for the entire heat exchange surface area A in:

$$\phi_q = \frac{1}{A}\frac{dQ}{dt} = -\lambda \frac{d\theta}{dx} \quad [1.3]$$

$\frac{dQ}{dt}$, heat flux or power, can both be noted $\dot{Q}$.

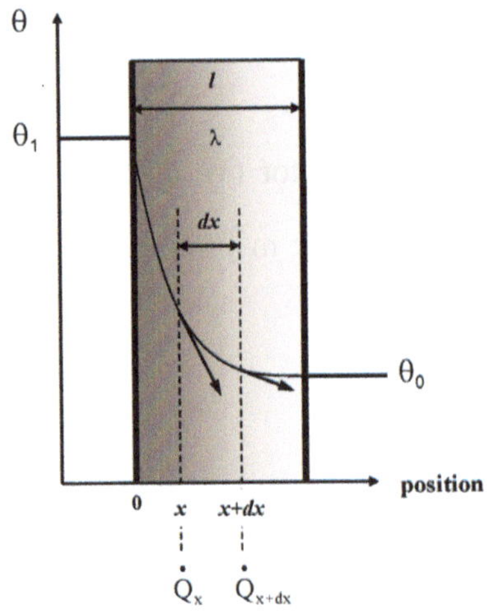

Figure 1.1. *Temperature profile in an infinite plate in a transient state. Temperature gradient is indicated by arrows at the x and x+dx sides*

If, starting from a uniform temperature θ_0 (Figure 1.1), one side of this plate is rapidly brought to the temperature θ_1, heat transfer takes place under the influence of the temperature gradient in a direction normal to the plate. After a certain time, the temperature profile is that shown as in Figure 1.1. From equation [1.3], the quantity of heat dQ that enters a section of thickness dx in a time interval dt through the constant-temperature surface at the x side is evidently greater than that leaving through the constant-temperature surface at the x+dx side. Heat therefore accumulates in this section, which results in a temperature increase in the plate element. This continues until that which enters x is equal to that which leaves x+dx, that is until $\frac{d\theta}{dx}$ is the same for all constant-temperature surfaces or until the temperature profile in the plate is linear (Figure 1.2).

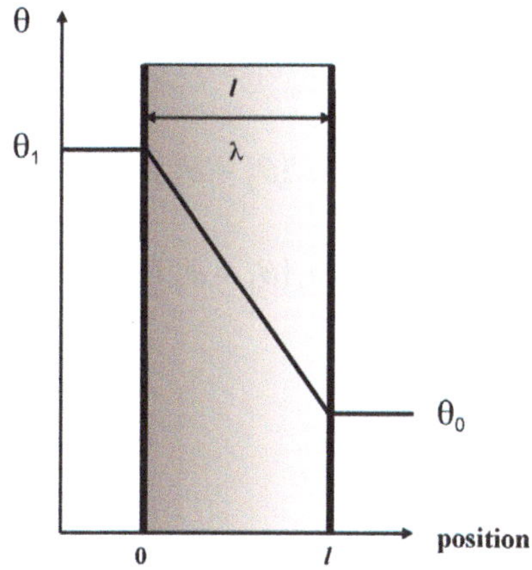

Figure 1.2. *Temperature profile in an infinite plate in a steady state*

1.1.1.2. *Heat transfer in an infinite plate in a steady state*

As we have just seen, heat transfer in a steady state infinite plate is characterized by a linear temperature gradient (Figure 1.2). Equation [1.3] is integrated on the plate thickness l in the following way:

$$\dot{Q} = A\ h\ \Delta\theta \qquad [1.4]$$

with $\Delta\theta = \theta_1 - \theta_0 > 0$ and:

$$h = \frac{\lambda}{l} \qquad [1.5]$$

h, the heat transfer coefficient (W m^{-2} K^{-1}), characterises the ability of the plate to allow heat to pass through since it incorporates both thermal conductivity λ and thickness l.

1.1.1.3. *Heat transfer in a steady state composite plate*

A wall is often composed of a several barriers to the heat transfer: the heat exchange plate itself, fouling deposits, boundary layers of fluid in contact with the surfaces, etc. The composite plate model is useful therefore, given its multiple applications.

In a steady state, the heat flux density reaches a constant value overall and in each of the sub-plates comprising the composite plate (Figure 1.3):

$$\phi_q = \frac{1}{A}\frac{dQ}{dt} = \begin{cases} \lambda_1 \dfrac{\theta_1 - \theta_2}{l_1} = h_1(\theta_1 - \theta_2) \\ \lambda_2 \dfrac{\theta_2 - \theta_3}{l_2} = h_2(\theta_2 - \theta_3) \\ \lambda_3 \dfrac{\theta_3 - \theta_4}{l_3} = h_3(\theta_3 - \theta_4) \end{cases} \qquad [1.6]$$

This is equivalent to:

$$\Leftrightarrow \phi_q = \frac{1}{\dfrac{1}{h_1} + \dfrac{1}{h_2} + \dfrac{1}{h_3}} \Delta\theta \qquad [1.7]$$

The heat flux density is therefore proportional to the temperature difference and the transfer coefficient relating to all sub-plates, denoted by h_g (W m^{-2} K^{-1}), also known as the overall heat transfer coefficient, which is as follows:

$$\begin{cases} \dfrac{1}{h_g} = \dfrac{1}{h_1} + \dfrac{1}{h_2} + \dfrac{1}{h_3} \\ \phi_q = h_g \Delta\theta \end{cases} \qquad [1.8]$$

More generally, for a stack of successive i layers (thickness l_i, thermal conductivity λ_i):

$$\frac{1}{h_g} = \sum_i \frac{1}{h_i} \text{ with } \frac{1}{h_i} = \frac{l_i}{\lambda_i} \qquad [1.9]$$

The terms $\frac{1}{h}=\frac{l}{\lambda}$ are referred to as thermal resistance. In fact, by analogy with electrical resistance, the overall thermal resistance of a system is the sum of the individual thermal resistances that constitute the system.

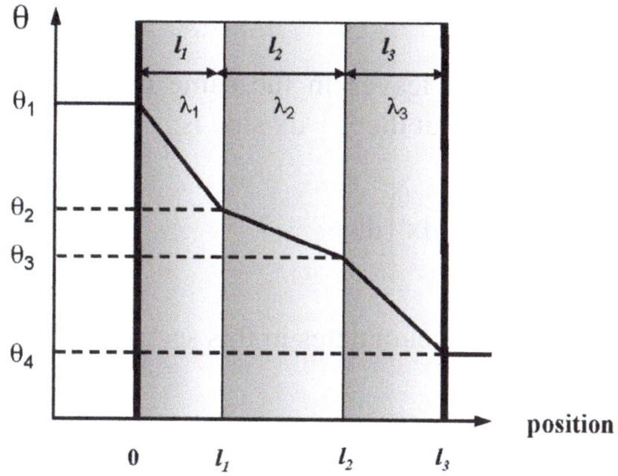

Figure 1.3. *Temperature profile in a succession of infinite plates joined together in a steady state*

1.1.1.4. *Heat transfer in a non-steady state*

In the study of heat transfer, the transient phase is often neglected, focusing mainly on unit processing equipment (pasteurizers, sterilizers, vacuum evaporators etc.) under steady state conditions. However, it is sometimes necessary to study transfer processes under transient state (dynamic) conditions. An illustrative example is the calculation of sterilization rates, which are based on temperature change at the core of a product over time (see Chapter 4).

To simplify our approach, we will keep the geometric model of the semi-infinite plate (Figure 1.1). The initial conditions are:

– the entire plate is at a uniform temperature θ_0;

– at 0 time, one side is rapidly brought to temperature θ_1, which is then assumed to be constant over time, and the other side is maintained at θ_0.

To evaluate heat transfer, we simply apply the qualitative description in section 1.1.1.1. The amount of heat that enters a section of thickness dx in a time interval dt through the constant-temperature surface at the x side is:

$$-\lambda\,\mathrm{A}\frac{\mathrm{d}\theta}{\mathrm{dx}}\,\mathrm{dt}$$

The amount of heat that leaves in the same time interval through the constant-temperature surface at the x + dx side is:

$$-\lambda\,\mathrm{A}\left(\frac{\mathrm{d}\theta}{\mathrm{dx}}+\frac{\mathrm{d}}{\mathrm{dx}}\left(\frac{\mathrm{d}\theta}{\mathrm{dx}}\right)\mathrm{dx}\right)\mathrm{dt}$$

The amount of heat that accumulates in this section is therefore:

$$\lambda\,\mathrm{A}\frac{\mathrm{d}^2\theta}{\mathrm{dx}^2}\,\mathrm{dx\,dt} \qquad [1.10]$$

In addition, this accumulation of heat in the section of thickness dx, volume dV = A·dx, and mass dm = A·dx·ρ, results in a temperature variation dθ, which satisfies the relationship:

$$\mathrm{dm}\,C_p\,\mathrm{d}\theta = \mathrm{A\,dx}\,\rho\,C_p\,\mathrm{d}\theta \qquad [1.11]$$

where C_p (J kg^{-1} K^{-1}) is the specific heat capacity of the product. Hence, by equating expressions [1.10] and [1.11]:

$$\frac{\mathrm{d}\theta}{\mathrm{dt}}=\frac{\lambda}{\rho\,C_p}\frac{\mathrm{d}^2\theta}{\mathrm{dx}^2}=\mathrm{D_q}\frac{\mathrm{d}^2\theta}{\mathrm{dx}^2} \qquad [1.12]$$

$$\text{with } \mathrm{D_q}=\frac{\lambda}{\rho\,C_\mathrm{p}} \qquad [1.13]$$

D_q is the thermal diffusivity (m^2 s^{-1}).

This equation provides analytical solutions for simple geometric shapes such as infinite plates, infinite cylinders or spheres. Resulting charts allow the calculation of heat transfers and temperature changes, and show dimensionless temperature θ* (so-called reduced temperature, representing the transient temperature distribution) as a function of the Fourier number Fo at a given position of the shape in question (Figures 1.4, 1.5 and 1.6). θ* and Fo are functions of temperature and time, expressed non-dimensionally:

$$\theta^* = \frac{\theta_1 - \theta}{\theta_1 - \theta_0} \qquad [1.14]$$

$$\mathrm{Fo} = \frac{\lambda}{\rho\, C_p} \frac{t}{l^2} \qquad [1.15]$$

where l is a characteristic length of a given geometric shape. It corresponds to the half-thickness of an infinite plate or to the radius of an infinite cylinder or sphere, commonly denoted as Δx and R_{max}, respectively. Note that θ* ranges between 1 (t_0) and 0 ($t\infty$), regardless of the values of θ_0 and θ_1.

Using these graphs, it is possible to determine the change in temperature in finite solids formed by invoking these infinite shapes (Newman expression). For example, a point in a food may be visualized as the intersection of an infinite plate and an infinite cylinder. In this case, the dimensionless temperature at a point in the finite solid is obtained by multiplying the dimensionless temperatures at the same point in the infinite solids that comprise it.

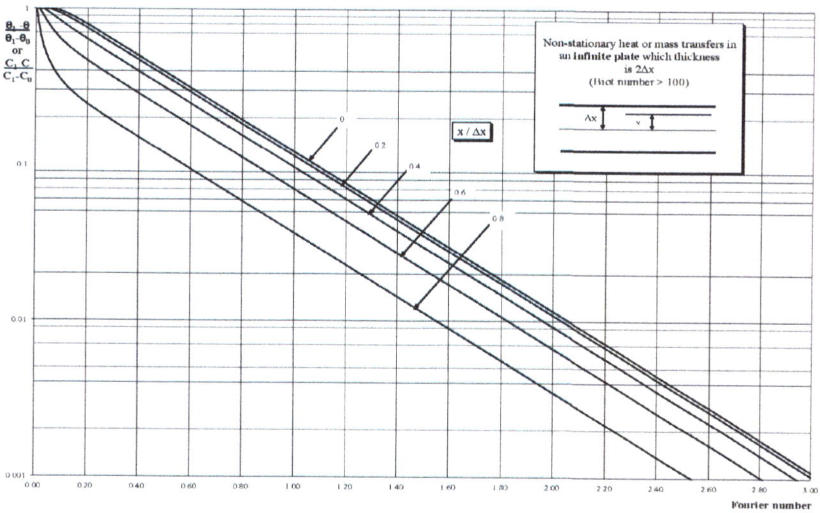

Figure 1.4. *Non-stationary heat or mass transfers in an infinite plate*

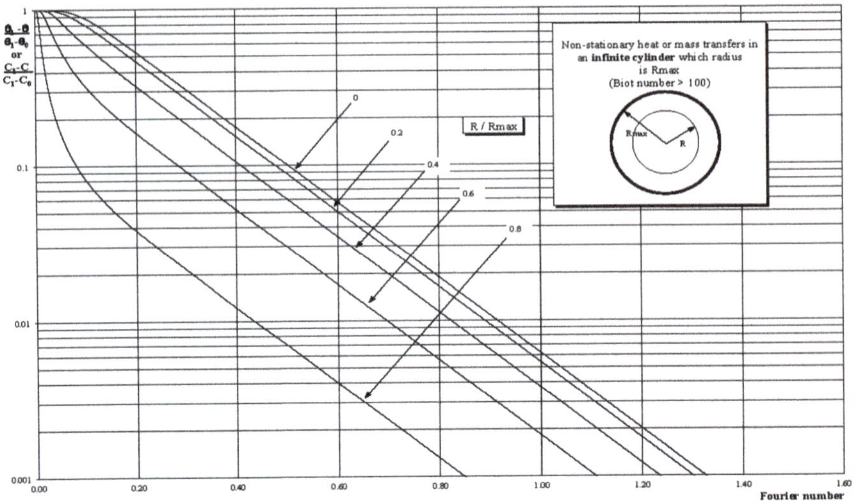

Figure 1.5. *Non-stationary heat or mass transfers in an infinite cylinder*

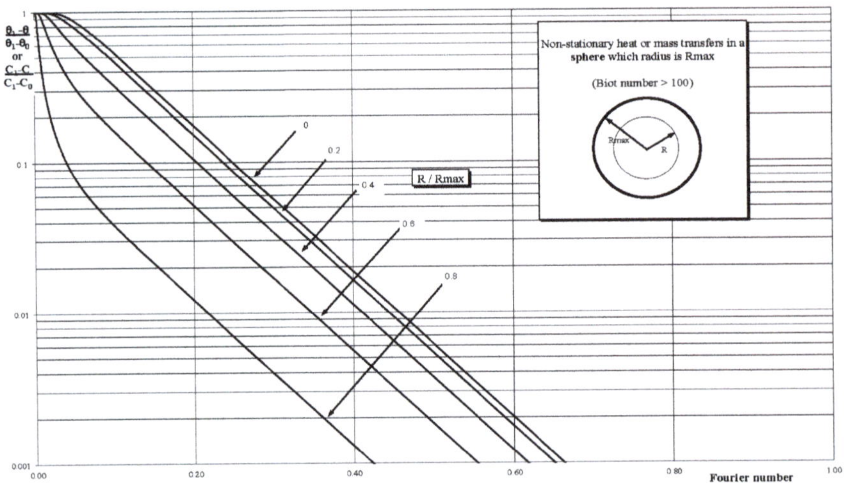

Figure 1.6. *Non-stationary heat or mass transfers in a sphere*

Time and space discretization methods offer more general solutions to solving this equation. Bimbenet and Loncin [BIM 95] provide a very good overview of these methods.

1.1.2. *Mass transfer: Fick's law*

Fick's law is expressed by the vector relationship:

$$\vec{\phi}_{m} = -D_{m}\,\overrightarrow{\text{grad}(C)} \qquad [1.16]$$

where $\vec{\phi}_m$ is the mass flux density vector (kg s^{-1} m^{-2}), i.e. the scalar value $\frac{1}{dA}\frac{dm}{dt}$ (dA: elementary surface exchange; m^2) multiplied by the normal unit vector at a constant-temperature surface:

$$\varphi_{m} = \frac{1}{dA}\frac{dm}{dt} = -D_{m}\frac{dC}{dx} \qquad [1.17]$$

The proportionality factor D_m between the mass flux density and the concentration gradient (kg m^{-3}) is the diffusion coefficient (m^2 s^{-1}), characteristic of a given product.

There is a similarity between equations [1.16, 1.17] and equations [1.1, 1.2]: Fourrier's law describes heat exchange due to a temperature gradient, and Fick's law the change in mass due to a concentration gradient. In other words, Q can be replaced by m, λ by D_m and θ by *C* to switch from Fourrier's law to Fick's law.

As a result, Fourrier's law and Fick's law are interchangeable for transient and steady states for a given geometry. In the non-stationary state, the change in concentration, at a given point, as a function of time is:

$$\frac{dC}{dt} = D_{m}\frac{d^{2}C}{dx^{2}} \qquad [1.18]$$

An identical approach to that proposed for heat transfers can provide solutions to this equation (see section 1.1.4): Figures 1.4, 1.5 and 1.6 give reduced concentration (C*) at a given time (Fo value) and position for each of the geometric shapes.

In the case of the semi-infinite plate (thickness *l*) and under steady state conditions, integration of [1.16] results in:

$$\dot{m} = A\,k\,\Delta C \qquad [1.19]$$

with $\Delta C > 0$ and:

$$k = \frac{D_m}{l} \qquad [1.20]$$

k being the mass transfer coefficient (m s^{-1}).

1.1.3. *Momentum transfer*

In this case, the concept of conduction is comparable to the transport of momentum by friction between parallel layers: this momentum transfer occurs perpendicularly to the flow direction, excluding the random transfer of liquid elements in this direction. This type of flow, known as laminar flow, occurs for example in a cylindrical tube: each ring-shaped liquid layer rubs against the two adjacent layers, and the momentum transfer takes place from the centre of the pipe to the walls.

Laminar flow is therefore fundamentally different to turbulent flow, which is characterized by a chaotic motion of fluid elements.

1.1.3.1. *Newton's law*

The model that underlies this law consists of two parallel plates, one fixed and the other separated by a distance of dx, with an area A and moving at a constant velocity under the influence of a tangential force F (Figure 1.7). It induces the laminar flow of a fluid present between these plates.

By analogy with previous discussion on heat and mass transfer, this model can also be represented in the steady state by a constant shear rate between the two plates (Figure 1.7; ❷).

Newton's law is expressed by the vector relationship :

$$\overrightarrow{\varphi_{mv}} = -\eta \, \overrightarrow{\text{grad}(v)} \qquad [1.21]$$

$\overrightarrow{\phi_{mv}}$ is the momentum flux density vector (kg m^{-1} s^{-2}), i.e. the scalar value $\frac{1}{dA}\frac{d(m\,v)}{dt}$ (dA: elementary exchange surface) multiplied by the normal unit vector at the isovelocity surface:

$$\varphi_{mv} = \frac{1}{dA}\frac{d(m\ v)}{dt} = -\eta\frac{dv}{dx}. \qquad [1.22]$$

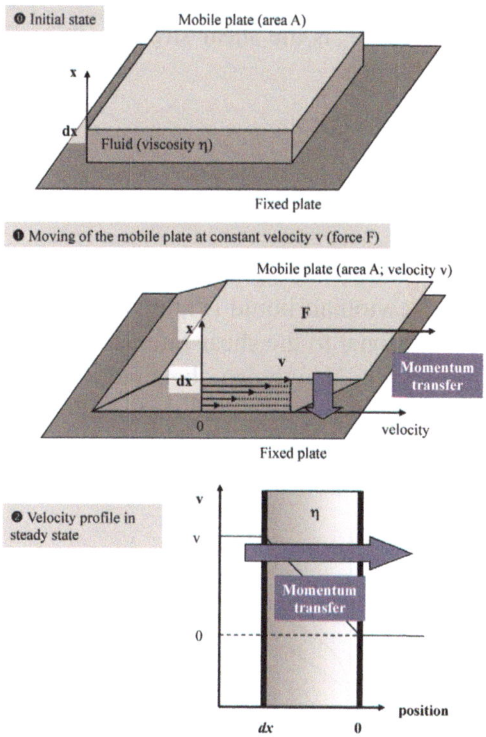

Figure 1.7. *Newton's model and velocity profile in an infinite plate in a steady state*

The proportionality factor η (Pa s) is defined as the dynamic viscosity.

Given that:

$$\frac{d(m\ v)}{dt} = m\frac{dv}{dt} = m\ a = F$$

Using equation [1.22], it is therefore possible to explain the force involved in the flow of a liquid in laminar state. This force is proportional to the surface A and to the shear rate (or velocity gradient) $\frac{dv}{dx}$ (s^{-1}):

$$F = -A\ \eta\ \frac{dv}{dx} \qquad [1.23]$$

The force per unit area $\frac{F}{A}$ is the shear stress τ (Pa) and the shear rate is typically denoted as $\dot{\gamma}$ (s^{-1}):

$$\tau = -\eta \dot{\gamma} \qquad [1.24]$$

Newton's model assumes that η is constant regardless of shear rate, which characterises a Newtonian liquid (Figure 1.8 – curve 1): in this case, the shear stress is proportional to the shear rate. However, in the food sector, many liquids display non-Newtonian behaviour. The study of such behaviour comes under rheology, a science that studies the deformation of a body based on the stress applied to it. Some typical fluid models are described in Figure 10, and explained in more detail in Volume 1 [JEA 16a].

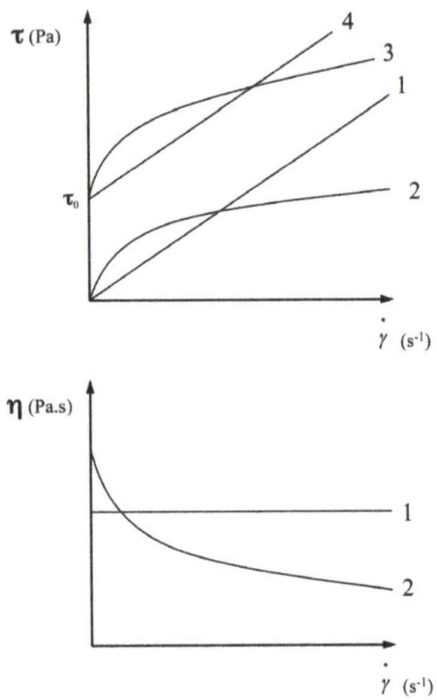

Figure 1.8. *Rheological models*

If viscosity decreases with shear rate, the behaviour is known as shear thinning (Figure 1.8 – curve 2). Materials in which deformation occurs above the yield stress τ_o (Figure 1.8 – curves 3 and 4) demonstrate viscoplastic behaviour. The model described by curve 4, which displays Newtonian behaviour beyond τ_o applies to a "Bingham plastic" material.

$$\tau - \tau_o = -\eta \frac{\mathrm{dv}}{\mathrm{dx}} \qquad [1.25]$$

1.1.3.2. *Calculation of laminar flow: Poiseuille's law*

Knowing the relationships that describe the flow of liquids in different types of ducts is necessary to measure fluid transfer through pipe sections, pumps, etc. In addition, these relationships apply to control systems fundamental to the optimization of process flow. For example, consider the case of laminar flow in a cylindrical pipe. If we isolate a coaxial cylindrical liquid element moving through this pipe, the sum of forces on this element yields the following results:

$$\begin{cases} \tau_x = \dfrac{\Delta P}{L} \dfrac{x}{2} \\ \tau_0 = 0 \\ \tau_w = \dfrac{\Delta P}{L} \dfrac{R}{2} \end{cases} \qquad [1.26]$$

These relationships apply as long as flow is laminar in the gap between the pipe wall and the cylindrical liquid element, even if the flow inside it is turbulent. As before, by combining equations [1.24] and [1.26], we obtain:

$$\mathrm{dv} = -\frac{\Delta P}{2\,\eta\,L}\,\mathrm{x}\,\mathrm{dx}$$

This expression integrates from x to R, with maximum velocity at x = 0 (v_{max}), equal to v_x at x and zero at the wall ($v_R = 0$):

$$\begin{cases} v_x = \dfrac{\Delta P}{4\,\eta\,L}(R^2 - x^2) \\ v_{max} = \dfrac{\Delta P}{4\,\eta\,L} R^2 \end{cases} \qquad [1.27]$$

Figure 1.9 gives v and τ as a function of position x. According to [1.26], τ is maximum at the wall (x = ± R) and zero at x = 0.

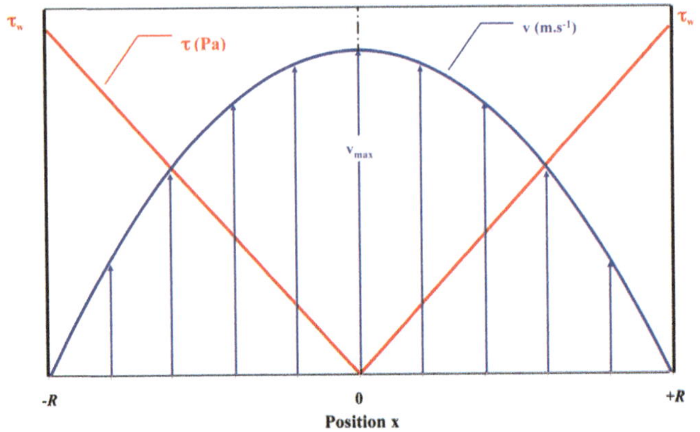

Figure 1.9. *Velocity v and shear stress τ profiles for laminar flow in a cylindrical pipe*

Using the expression of local velocity, it is possible to determine the volumetric flow rate $\dot{V}$ and the average velocity $\bar{v}$ of the liquid flowing in the pipe:

$$\dot{V} = \frac{\Delta P}{8\ \eta\ L} \pi\ R^4 \qquad [1.28]$$

$$\bar{v} = \frac{\Delta P}{8\ \eta\ L} R^2 = \frac{v_{max}}{2} \qquad [1.29]$$

1.2. Convective transfer processes

1.2.1. *Introduction to geometric and physical similarity: Reynolds' experience*

The work of Osborne Reynolds, the English physicist (1842–1912), focused mainly on the study and characterization of liquid flow in cylindrical pipes. In 1883, he visualized the flow behaviour of water in cylindrical glass tubes using a dye (Figure 1.10). He noted that the transition between laminar and turbulent flow under various experimental conditions (fluid velocity v, pipe diameter D, fluid viscosity and density respectively η and ρ) systematically occurs at the same value for the ratio $\frac{D \rho \text{v}}{\eta}$, that is approximately equal to 2,000, now referred to as the Reynolds number (Re). It is the ratio of inertial forces to viscous forces (thus dimensionless), and helps characterise flow behaviour. This number forms the basis of studies on geometric and physical similarities between systems: the tubes studied by Reynolds were all cylindrical and assumed to be infinite (small diameter compared to the length), and flow behaviour was the same for identical ratios of inertial to viscous forces. This kind of ratio makes it possible to easily scale up the systems.

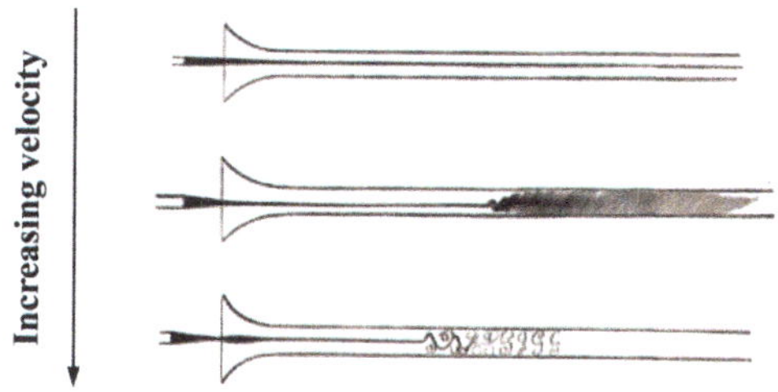

Figure 1.10. *Reynolds' experiment*

1.2.2. *Importance of similarity*

It is easy to obtain a description of physical phenomena during heat transfer by conduction, since the basic laws are known. However, in the case of heat transfer by convection, these laws no longer apply, and with the current state of knowledge, it is not possible to predict the behaviour of a given system or calculate its characteristic values.

Under these conditions, predicting the behaviour of physical systems is based on experimental results, according to two very similar approaches:

– model analysis involves developing a normally reduced model of a given system, similar geometrically and as similar as possible physically, and extrapolating the results, expressed as a relationship between their physical quantities, from the model to the system;

– a study of common geometrically-defined systems (infinite cylindrical pipe, infinite plane wall, sphere, etc.) allows extrapolation of the relationships between their physical quantities to any geometrically similar system.

It is clear, therefore, that geometric similarity is fundamental to the study of physical phenomena. It is particularly important to characterize the geometric scale-up of two similar systems by the relationships between homologous lengths.

For example, a characteristic length commonly used for fluid conduits is the hydraulic diameter (D_h) defined by the relationship:

$$D_h = \frac{4 \times \text{cross} - \text{sectional area}}{\text{wetted perimeter}} \quad [1.30]$$

where the wetted perimeter is the perimeter of the pipe cross-section in contact with the fluid. The expression of D_h therefore depends on the geometry of the pipe:

– for a cylindrical pipe of radius R, $D_h = 2R$;

– for a rectangular pipe (width of plates l, separated by a distance e, which is negligible compared to l), $D_h = 2e$.

Note that the critical value of the Reynolds' number 2,000 is valid in the case of cylindrical pipes, but is entirely dependent on the geometry. For example, in the case of rectangular pipes (e.g. plate heat exchangers), this value is usually much lower, at around 200.

1.2.3. *Complete similarity*

Geometric similarity implies that the ratios of homologous lengths of two systems are constant (Figure 1.11):

$$\frac{l'_a}{l_a} = \frac{l'_b}{l_b} = l_+ \quad [1.31]$$

It requires scaling consistency of both surfaces ($A_+=l_+^2$) and volumes ($V_+=l_+^3$).

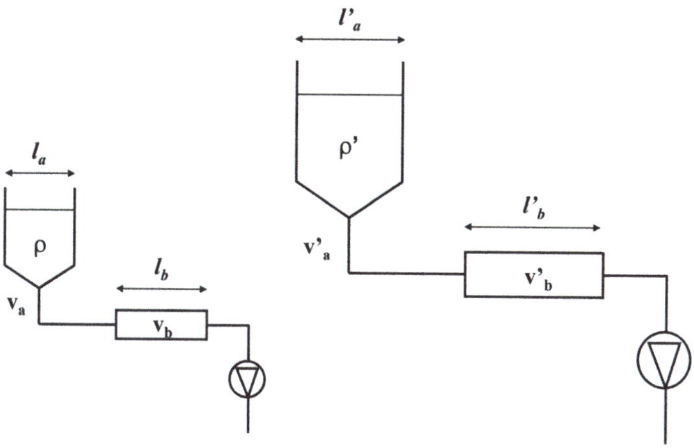

Figure 1.11. *Similar systems*

Physical similarities can be defined based on the same principle. For example, in the case of Figure 1.11, physical similarity of velocities is achieved once the velocity ratio is constant at all homologous points:

$$\frac{v'_a}{v_a} = \frac{v'_b}{v_b} = v_+ \qquad [1.32]$$

The fluids used in both systems may differ, in particular with regard to their densities. In this case, the scaling of densities is given by:

$$\frac{\rho'}{\rho} = \rho_+ \qquad [1.33]$$

which implies a scaling of masses:

$$m_+ = \rho_+ V_+ \qquad [1.34]$$

Bimbenet and Loncin (Table 1.1) point out that choosing scaling factors of lengths l_+, velocities v_+, and densities ρ_+ respectively equal to 10, 5 and 1 result in the scaling of the dynamic viscosities η_+ equal to 50, and the scaling of surface tensions σ_+ equal to 250. Although it may be possible to increase viscosity by a factor of 50 using a thickener, surface tension in fluids is not known to differ by a factor of 250.

Physical quantities	*Scaling*	*Numerical example*
Lengths	l_+	10
Velocities	v_+	5
Densities	ρ_+	1
Surfaces	$A_+ = l_+^2$	100
Volumes	$V_+ = l_+^3$	1,000
Accelerations	$a_+ = v_+^2/l_+$	2.5
Masses	$m_+ = V_+ \rho_+$	1,000
Inertial forces	$F_+ = m_+ a_+$	2,500
Viscosities	$\eta_+ = l_+ v_+ \rho_+$	50
Surface tensions	$\sigma_+ = l_+ v_+^2 \rho_+$	250

Table 1.1. *Examples of scaling of physical quantities in two similar systems [BIM 95]*

Systems of exact similarity effectively do not exist and we must remain satisfied with partial similarities appropriate for the transport phenomena studied. For example, the study of the relationships between inertial and viscous forces carried out by Reynolds would be meaningless in a capillary tube where surface tension forces dominate.

1.2.4. *Partial similarity*

1.2.4.1. *Dimensionless numbers in fluid dynamics*

In fluid dynamics, the systems studied are assumed to be similar in terms of inertial forces, which in turn serve as a reference for other forces.

The relationship between different types of forces leads to the use of dimensionless numbers, also referred to as similarity invariants. The value of a similarity invariant determines the relative importance of both types of forces involved in its calculation.

The ratio of pressure forces ($F_p = P\ l^2$) to inertial forces ($F_i = l^2\ v^2\ \rho$) results in a dimensionless number called the Euler number (Eu):

$$\mathrm{Eu} = \frac{P}{v^2\ \rho} \qquad [1.35]$$

Likewise, the dimensionless Froude number (Fr) makes it possible to assess the similarity of different systems with respect to gravitational forces ($F_g = l^3\ \rho_g$) and inertial forces:

$$\mathrm{Fr} = \frac{v^2}{l\ g}. \qquad [1.36]$$

As already seen, the Reynolds number is the ratio of inertial forces to viscous forces ($F_v = A\ \eta \frac{v}{l} = l\ \eta\ v$; equation [1.23]):

$$\mathrm{Re} = \frac{l\ \rho\ v}{\eta} \qquad [1.37]$$

Likewise, the dimensionless Weber number (We) makes it possible to assess the similarity of different systems with respect to surface tension and inertial forces:

$$\mathrm{We} = \frac{l\ \rho\ v^2}{\sigma} \qquad [1.38]$$

1.2.4.2. *Dimensionless numbers and heat transfer*

As in the case of fluid dynamics, it is possible to define dimensionless numbers to study heat transfer phenomena. It is generally accepted that convective heat transfer can be described by the following model (Figure 1.12):

– within a liquid (or "turbulent core"), convection is such that the temperature can be considered constant at all points;

– due to the friction between fluid and wall there is a layer of liquid at the wall called the “boundary layer” in which fluid flow is laminar. The thickness of this layer, which is dependent on the geometry of the system, fluid characteristics and velocity, is statistically equal to a constant value *l*, this statistic unfortunately being poorly understood. As a result, λ and *l* are often combined in the expression of the local heat coefficient α according to expression [1.5].

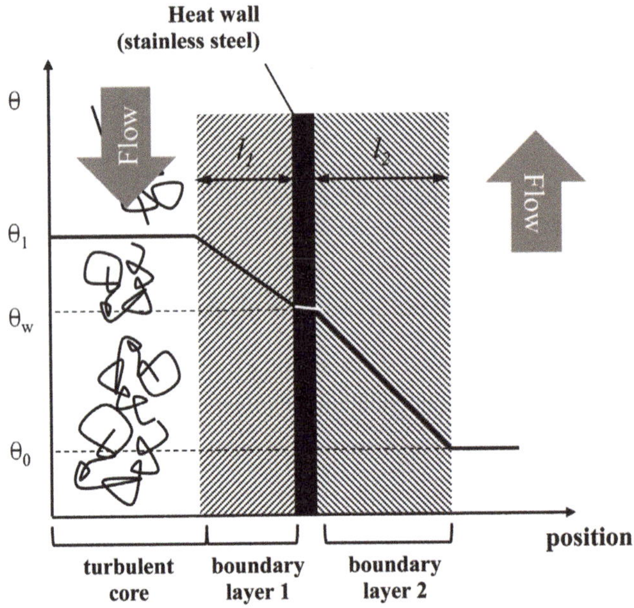

Figure 1.12. *Heat transfer in a boundary layer*

Of the various dimensionless numbers commonly used in convective heat transfer, the Nusselt (Nu) and Prandtl (Pr) numbers describe heat transfer between the mass of a fluid (viscosity η, thermal conductivity λ, specific heat capacity Cp) and a wall through a boundary layer (thickness l):

$$\mathrm{Nu} = \frac{h\,l}{\lambda} \qquad [1.39]$$

$$\mathrm{Pr} = \frac{\eta\,C_p}{\lambda} \qquad [1.40]$$

1.2.5. *Dimensional analysis*

1.2.5.1. *π theorem*

Many physical phenomena can only be described by equations, due to a lack of knowledge of the fundamental elements governing them. However, it is often possible to identify factors that affect these phenomena. Through experimentation, we can then establish the links between these factors.

Dimensional analysis reduces experimentation by grouping all variables into less dimensionless numbers, the links therefore being expressed by the relationships between these numbers. This analysis is based on the following assumption: relationships that link physical phenomena are dimensionally homogenous and are identical regardless of the coherent system of units used.

This assumption leads to the π-theorem, also known as the Vashy–Buckingham theorem:

> "*Any physical quantity Q (target parameter), that is a function of p independent dimensional variables measured by q fundamental units, can be described by an implicit function of [p – q + 1] dimensionless numbers.*"

In the following section, some examples of applications of this theorem will be given for fluid dynamics and heat transfer.

1.2.5.2. *Movement of a sphere in a fluid under the action of a force*

A sphere of diameter *D* moves at a constant velocity v in a fluid of density ρ and viscosity η (Figure 1.13). Under these conditions, the driving force F (target parameter) is equal to the resistive force, F, which is a function of D, ρ, v and η.

An experimental study of F, determined by the reciprocal action of four variables *D*, ρ, v and η, with ten values for each variable would result in 10^5 trials; such a study is obviously unrealistic.

π-theorem limits the number of variables by establishing a relationship between two dimensionless numbers. F is a function of four independent dimensional quantities measured by three fundamental units (length, mass and time), and can therefore be described by [4 – 3 + 1 = 2] dimensionless numbers.

These numbers are Re and the Newton number (Ne), from which the Euler number is derived:

$$\mathrm{Ne} = \frac{\mathrm{F}}{l^2\, \mathrm{v}^2\, \rho} \qquad [1.41]$$

Ne expresses the similarity between two systems in terms of inertial forces. l, the characteristic length of the systems for both Ne and Re, is in this case, the diameter of the sphere D, and Ne is adapted by using the cross-sectional area $\pi \frac{D^2}{4}$ and $\frac{\mathrm{v}^2}{2}$ instead of l^2 and v^2. This leads to a modified Newton number, Ne':

$$\mathrm{Ne'} = \frac{\mathrm{F}}{\pi \frac{D^2}{4} \frac{\mathrm{v}^2}{2} \rho} \qquad [1.42]$$

The following relationships are established through experimentation:

$$10^{-3} < \mathrm{Re} < 2 \quad \mathrm{Ne'} = 24(\mathrm{Re}^{-1}) \qquad [1.43]$$

$$2 < \mathrm{Re} < 5(10^2) \quad \mathrm{Ne'} = 18.5\,(\mathrm{Re}^{-0.6}) \qquad [1.44]$$

$$5(10^2) < \mathrm{Re} < 10^5 \quad \mathrm{Ne'} = 0.45 \qquad [1.45]$$

It is important to note that the relationships [1.43–1.45] linking Ne′ and Re are independent of the cause of movement, and reflect only the relationships linking F and v as a function of fluid characteristics and sphere diameter.

If the driving force F is linked solely to gravity, and expressed as the resulting force of the weight of the sphere and buoyancy, we obtain Stokes' equation [1.1] for Re < 2.

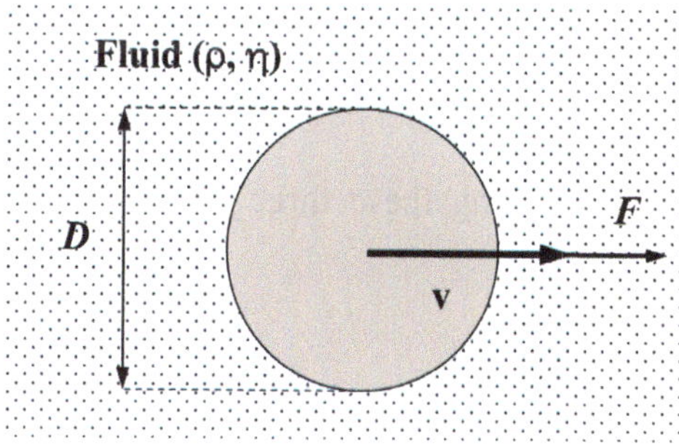

Figure 1.13. *Movement of a sphere in a fluid*

1.2.5.3. *Pressure drop in a cylindrical pipe*

The pressure drop $\Delta P = P_1 - P_2$ (Figure 1.14) in a circular pipe of diameter D and length L depends on the average velocity $\bar{v}$ of the fluid, its viscosity η and density ρ. ΔP (target parameter) is thus a function of five-dimensional independent quantities (L, D, $\bar{v}$, η and ρ) measured by three fundamental units (length, mass and time), and can be described as a function of [5 – 3 + 1 = 3] dimensionless numbers.

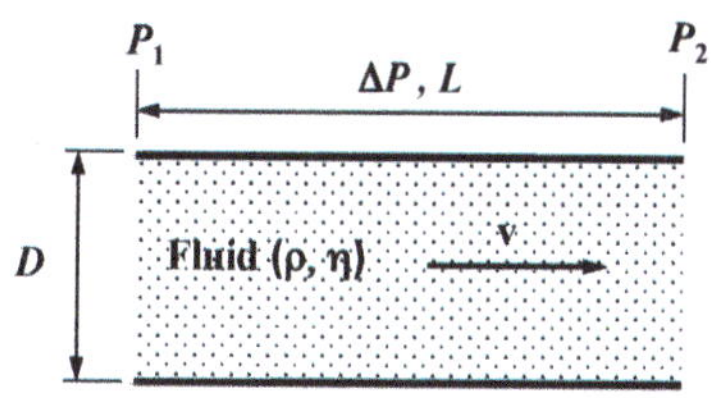

Figure 1.14. *Pressure drop in a cylindrical pipe*

These three invariants are generally Re, Eu and $\frac{L}{D}$, a dimensionless number relating to the geometry of the pipe. Re and Eu are adapted to each particular case as follows:

$$\mathrm{Re} = \frac{D \rho \bar{v}}{\eta} \qquad [1.46]$$

$$\mathrm{Eu} = \frac{2\,\Delta P}{\bar{\mathrm{v}}^2 \rho} \qquad [1.47]$$

The relationship f_1 linking these three dimensionless numbers can be written as follows:

$$\mathrm{Eu} = f_1\left(\frac{L}{D}, \mathrm{Re}\right)$$

or, by substituting Eu by its value and showing ΔP:

$$\Delta P = \frac{\bar{\mathrm{v}}^2 \rho}{2} f_1\left(\frac{L}{D}, \mathrm{Re}\right)$$

The pressure drop for an incompressible fluid is proportional to the length of the pipe: it is therefore possible to remove $\frac{L}{D}$ from $f_1\left(\frac{L}{D}, \mathrm{Re}\right)$ in the previous expression in order to keep the relationship dimensionally homogeneous. We get:

$$\Delta P = \frac{\bar{\mathrm{v}}^2 \rho}{2} \frac{L}{D} f_2(\mathrm{Re}) \qquad [1.48]$$

$f_2(\mathrm{Re})$, which is dimensionless, is known as the Darcy number (Da):

$$\mathrm{Da} = \frac{2\,D}{\bar{\mathrm{v}}^2 \rho} \frac{\Delta P}{L} = f_2(\mathrm{Re}) \qquad [1.49]$$

Determining ΔP therefore means establishing relationships between two invariants, Da and Re ($\mathrm{Da} = f_2(\mathrm{Re})$; equation [1.49]). The following relationships are established through experimentation:

– if $\mathrm{Re} < 2\ (10^3)$, and the flow is laminar, we obtain:

$$\mathrm{Da} = \frac{64}{\mathrm{Re}} \qquad [1.50]$$

resulting in Poiseuille's equation [1.29]. When the flow is laminar, density no longer appears in the Da–Re relationship since only viscous forces affect this phenomenon. The Poiseuille equation is used, for example, to describe filtration phenomena. In this case, the flow regime is laminar since the diameter of the membrane pores is small and the velocity of the fluid is low;

– if $2\,(10^3) < \text{Re} < 10^5$, the flow regime is turbulent and the Blasius relationship is normally applied for smooth pipes:

$$\text{Da} = 0.316\left(\text{Re}^{-0.25}\right) \qquad [1.53]$$

For rough pipes, l_a is taken into account, which is the average height of asperities, and an additional invariant is introduced: $\frac{l_a}{D}$ (relative roughness). Moody's diagram (not shown) gives the relationship between Re, Da and $\frac{l_a}{D}$. The graph in Figure 1.15 shows Da as a function of Re in the case of fluid flow in a smooth cylindrical pipe.

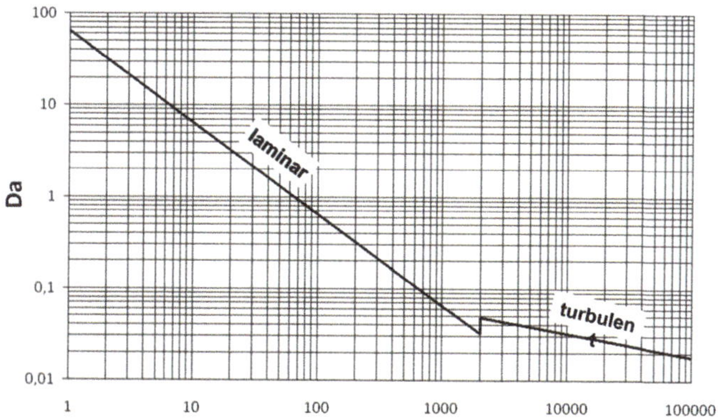

Figure 1.15. *Da – Re relationship for fluid flow in a smooth cylindrical pipe*

1.2.5.4. *Determination of heat transfer coefficients*

The local heat transfer coefficient α depends on:

– factors relating to the geometry of a given system (geometry is here characterized by a single dimension e.g. L);

– fluid flow conditions (characterized by average velocity $\bar{v}$);

– physical characteristics of the fluid (density ρ, specific heat C_p, thermal conductivity λ, viscosity η).

α (target parameter), a function of six independent dimensional quantities, expressed by four fundamental units (length, mass, time and temperature), can be described by a function of [6 – 4 + 1 = 3] dimensionless numbers. The invariants generally selected are Re, Nu and Pr. The relationship between the previous seven quantities can be summed up in the relationship Nu = *f* (Re, Pr) and the experimental data on a given geometry and flow type can be adjusted to an equation like:

$$Nu = C\ Re^{m}\ Pr^{n}$$

where C, m and n are constants. Here are some examples of specific situations.

Forced movement of fluid in a cylindrical pipe

For $8(10^{3}) < Re < 10^{6}$, $6(10^{-1}) < Pr < 5(10^{2})$ and $\frac{L}{D} > 50$:

$$Nu = 0.023\left(Re^{0.8}\right)\left(Pr^{0.33}\right)\left(\frac{\eta}{\eta_{w}}\right)^{0.14} \quad [1.52]$$

The characteristic length used for Nu and Re is the inner diameter (*D*) of the pipe. η is the dynamic viscosity in the turbulent core and η_W viscosity at the wall: the ratio $\frac{\eta}{\eta_w}$ is close to 1 for gas, but may be significant for liquids.

Forced movement of fluid parallel to a plane wall

For $Re < 10^{5}$

$$Nu = 0.664\left(Re^{0.5}\right)\left(Pr^{0.33}\right) \quad [1.53]$$

The characteristic length to take into account is the distance travelled by the fluid along the wall and where the velocity is the average velocity of the fluid $\bar{v}$.

Other relationships have been established for common heat transfer situations: fluidized beds, stirred tanks, heat transfer in free convection, heat transfers involving phase change (condensation, boiling), etc. However, most semi-empirical correlations between dimensionless numbers do not include the spatio-temporal variations of the physical properties of products during the process. Indeed, these correlations are usually constructed by considering a still matter with constant physical properties within the equipment, and more generally, throughout its entire transformation. Such approximations mean that the state of the system being studied is defined by a truncated (and insufficient) number of dimensionless numbers. This leads to non-generic process relationships since they are unable to describe the specific evolutions of the system resulting from the variability of the material physical properties. This is even more regrettable given the fact that physicochemical analysis techniques provide an ever finer characterization of the evolutions of the physical properties of matter subjected to complex stresses (temperature, shearing-time-deformation, pressure, etc.).

Delaplace *et al.* [DEL 15] give a deep insight on the theoretical framework which allows the principles of similarity theory to be respected in the case of processes using a material with constant or variable physical properties in the course of the transformation.

PART 2

Food Biological Stabilization

2

Inhibition of Food Modifying Agents

2.1. Refrigeration and freezing

2.1.1. *Definitions and basic principles*

2.1.1.1. *Definitions*

Lowering the temperature of a product can extend its shelf life and thereby delay consumption. Cold storage can be divided into two main categories:

– *refrigeration*: term used for low storage temperatures, above 0°C ("chilling temperature"). Refrigeration slows down enzymatic and chemical reactions, and consequently the growth and metabolism of microorganisms; however, it only ensures relatively short-term storage (few days). The growth of certain pathogens (*Listeria monocytogenes*, *Yersinia enterocolitica* and *Salmonella*) and spoilage bacteria (*Pseudomonas*, certain yeasts and fungi) is still possible at 6°C, but only *Clostridium botulinum* type E is still able to produce its neurotoxin below 10°C (see [JEA 16a], Chapter 3). Spoilage due to rancidity and fat hydrolysis may also during occur during chilled storage;

– *freezing*: general term for the phase transition of liquid water to ice, and maintaining the product at a sub-zero temperature. Freezing, therefore, combines the effects of a decrease in both temperature and a_w (see [JEA 16a], Table 2.4), since part of the water is in a crystalline state. The combination of these two effects means food can be stored for several

---

Chapter written by Romain JEANTET and Juliane FLOURY.

months at sub-zero temperatures. However, rancidity of unsaturated fatty acids remains problematic even at –20°C.

The term "deep-freezing" corresponds to the case where the product is frozen as quickly as possible at a temperature at or below –18°C, then kept at this temperature throughout storage.

2.1.1.2. *Influence of temperature on biochemical and biological reactions*

The reaction constant k_θ for a chemical, biochemical or enzymatic reaction depends on several parameters such as pH, water content, but in particular temperature. In most cases, the effect of temperature on the reaction rate constant can be described by the Arrhenius law:

$$k_\theta = a e^{\frac{-E}{RT}} \qquad [2.1]$$

where a is a constant (s^{-1}), E the activation energy of the reaction ($J\ mol^{-1}$), R the ideal gas constant ($8.314\ J\ mol^{-1}K^{-1}$) and T the absolute temperature (K). Equation [2.1] shows that a decrease in temperature causes a decrease in the rate constant k_θ, thus slowing down all the biochemical, enzymatic and chemical spoilage processes taking place in the food.

Some microorganisms may also die gradually when cells are placed in a very low-temperature atmosphere, usually at a logarithmic rate. This mortality is, however, still very marginal, and it is important to keep in mind that stabilization by freezing is primarily based on the inhibition and not the destruction of flora. As a result, the quality of the end product depends on the quality of the product prior to freezing, how quickly the product is cooled and frozen and whether the product is maintained at sub-zero temperatures during storage. After storage at a low temperature, microorganisms need to be exposed to a favorable temperature for a sufficiently long period before they can begin to reproduce and grow again, given its prior exposure to stress.

Psychrotrophic microorganisms can multiply actively at refrigeration temperatures (6°C). Psychrophilic microorganisms can continue to grow below –5°C and even at temperatures as low as –10°C. Most fungi and yeasts are psychrotrophic microorganisms and can therefore grow at normal refrigeration temperatures. Coliforms can also grow between 5°C and 0°C. In contrast, most pathogenic bacteria are unable to reproduce below 5°C and require this temperature or higher to be able to produce their toxin.

2.1.1.3. *Immobilization of water*

The water–ice phase transition that takes place during freezing has the advantage of fixing the tissue structure and separating the water fraction in the form of ice crystals. This water is therefore no longer available as a solvent or reactant. Consequently, the diffusion of solutes in the tissue is very slow, which contributes, together with a reduction in temperature, to a reduction in the rate of most reactions.

The formation of ice crystals has, however, the disadvantage of mechanically damaging the texture. When water turns into ice, the volume increases by approximately 9%. Subsequently, the ice undergoes a slight contraction upon further cooling. The constituents other than water, in particular lipids, contract during freezing. These volume variations, often non-homogeneous, cause internal stresses reaching several dozen MPa and are responsible for tearing and loss of cell contents (exudation) during thawing (as in the case of fruit, vegetables and meat).

2.1.2. *Ice formation*

2.1.2.1. *Nucleation and crystal growth*

Although the melting point of ice is 0°C at atmospheric pressure, the crystallisation of pure water begins at this temperature or below: the formation of ice crystals, also known as nucleation, is preceded by a phenomenon called "supercooling", which is more pronounced the faster the rate of cooling. In other words, water remains in a liquid state for a certain amount of time at a temperature below the melting point. Nucleation increases in the presence of insoluble salt crystals or solid particles such as dust. During the freezing of food products, nucleation is heterogeneous due to the presence of several solutes and supercooling is rarely observed.

The growth of ice crystals is controlled by two physical processes: a mass transfer during the diffusion of water molecules to ice crystals, and a heat transfer by diffusion of latent heat of crystallisation from the surface of the crystal to the liquid. The growth of ice crystals results from the migration of water molecules from the serum to existing nucleation seeds. It can therefore occur even at temperatures close to freezing point; in practice, the growth rate of ice crystals depends on the cooling rate of the product.

The size of ice crystals at the end of freezing depends on the number of seed crystals originally formed in the liquid medium. Therefore, regulating nucleation by temperature is essential in order to obtain crystals of a desired size. At low temperatures, nucleation is fast and many seeds produce a large number of small crystals, whereas at temperatures close to melting point, nucleation is slow, crystal nuclei are few in number, and the result is larger crystals. The shape of ice crystals strongly depends on the rate of crystal growth, and consequently on the freezing temperature. Rapid freezing produces almost spherical crystals, whereas slow freezing produces elongated, needle-shaped crystals.

In most systems that undergo freezing, the limiting factor is the internal heat transfer. In the case of very rapid freezing of fluids at very low temperatures, the limiting factor is mass transfer because the high viscosity of the liquid phase slows down the movement of water molecules. There is no phase transition below the glass transition temperature due to the lack of diffusion of molecules: the product is frozen in an amorphous state. This type of freezing is used in the cryopreservation of living cells (sperm, tissue, etc.) where the formation of ice crystals should be avoided as it may affect the cell structure. In this case, organic products (cryoprotectants) are added, which limit crystallisation by increasing viscosity and reducing the amount of freezable water.

2.1.2.2. *Freezing temperature of food*

The temperature changes in pure water and a dilute aqueous solution during freezing are shown in Figure 2.1.

In the case of pure water, ice crystals form and release heat after supercooling (point S in Figure 2.1). The temperature rises to 0°C and remains constant throughout crystallisation. When all the water has transformed into ice, the temperature decreases at a faster rate than at the beginning because the specific heat of ice is lower than that of water.

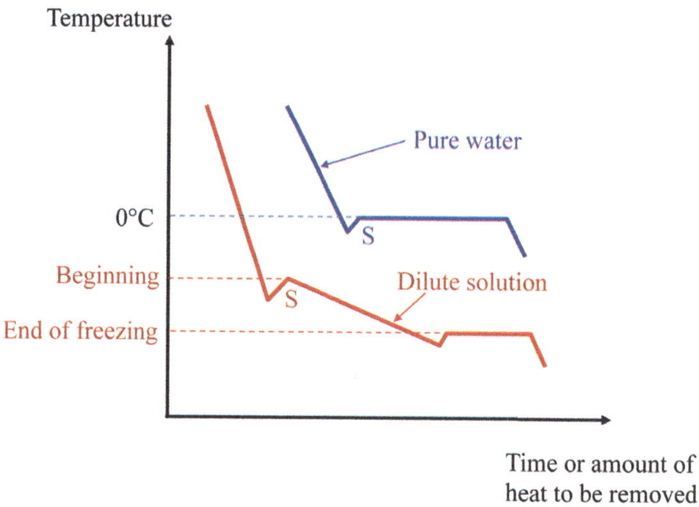

Figure 2.1. *Freezing curves of pure water and an aqueous solution*

In the case of a dilute solution, the temperature at the onset of freezing is less than 0°C. This cryoscopic lowering, noted $\Delta\theta$, is inversely proportional to the mole fraction of solutes in solution according to Raoult's law:

$$\Delta\theta = k\frac{m}{M} \quad [2.2]$$

with m being the concentration of solutes in solution (kg kg^{-1}), M their molar mass (kg mol^{-1}) and k the cryogenic constant of water (1.86°C kg^{-1} mol). During the crystallization of water, the freezing temperature gradually decreases due to an increase in the solute concentration in the residual aqueous phase.

Once a dissolved compound reaches its saturation point in the liquid phase, there is a simultaneous crystallization of ice and the solute. The mixture that crystallizes has a constant composition corresponding to the saturation concentration, and is called a eutectic mixture (Figure 2.2).

The temperature at the end of freezing, referred to as the eutectic or cryohydric temperature, is an invariable characteristic for a given solute, and remains constant throughout the crystallization of the eutectic mixture (Figure 2.2).

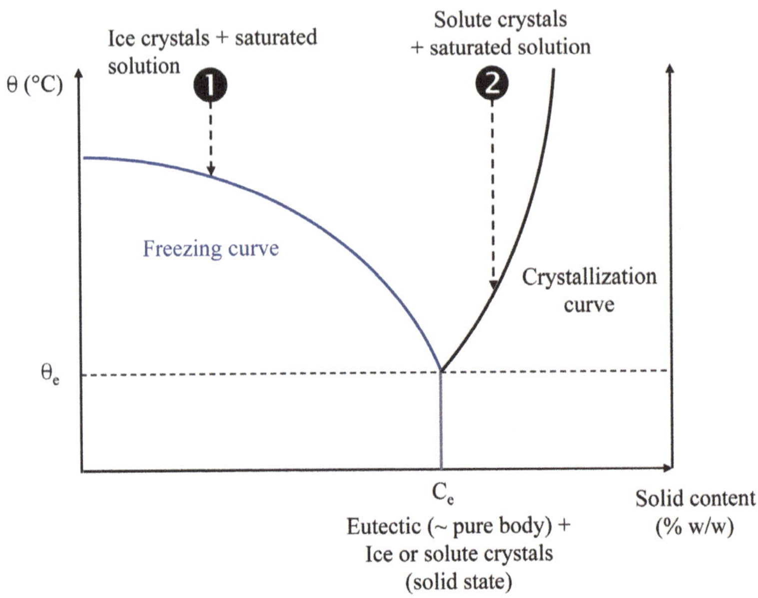

Figure 2.2. *Phase diagram of a simple binary mixture*

In the case of foods, there is rarely a straightforward eutectic transformation: the temperature at the end of freezing (total freezing of freezable water, or maximum solidification) of a food is usually below the eutectic temperature (Table 2.1).

Monitoring the temperature at the core of a food as a function of time during freezing gives quite regular curves, where the start of freezing and the various eutectics can generally not be determined (Figure 2.3). We will see later that the curves not only depend on the product characteristics but also on the rate of heat transfer.

Solution or food	Temperature	
	At start of freezing (°C)	At end of freezing (°C)
NaCl 1 M	–3.84	–21.13 (eutectic: 22.4% w/w)
Sucrose 1 M	–2.82	–9.50 (eutectic: 56.2% w/w)
Fish, meat (muscle), egg white	–0.60 to –2.20	–52 to –55
Fruit, vegetables	–0.90 to –2.70	
Milk	–0.52	
Bread	≈ –0.50	–70

Table 2.1. *Temperatures at the start and end of freezing (eutectic) of simple solutions and certain foods*

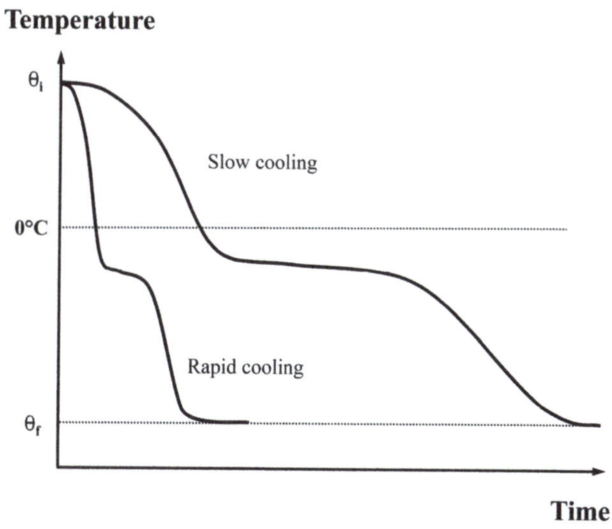

Figure 2.3. *Freezing curves of a real food. Comparison between slow and rapid cooling*

At the normal storage temperature of frozen foods (–18°C), a considerable proportion of freezable water is still in liquid state (2 to 15%) and has solvent and reactant properties (Figure 2.4). As a result, adverse changes can occur in frozen products due to the presence of residual liquid regions at high concentrations of solutes. It is therefore recommended to store frozen products at temperatures below –18°C. Moreover, in contrast to rapid freezing, slow freezing leads to the formation of very pure ice crystals, which can also cause freeze concentration of solutes in the residual liquid phase.

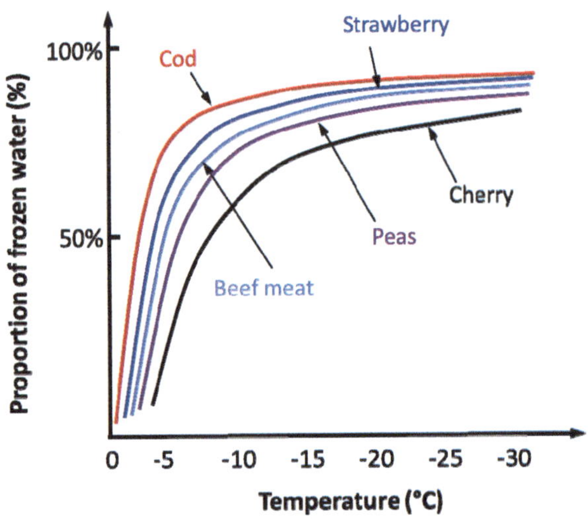

Figure 2.4. *Proportion of frozen water in different foods as a function of storage temperature*

2.1.3. *Freezing process*

2.1.3.1. *Freezing kinetics and energy use*

From a thermodynamic point of view, freezing a food product involves the removal of heat to lower the temperature (Figure 2.3).The required cooling capacity ($\dot{Q}$) is defined as the amount of heat to be removed from the food product Q_T per unit of time. This value is taken into account in the design of the freezing facility. The calculation of Q_T can be broken down as follows: the first step involves lowering the initial temperature θ_i (°C) of the food product to the initial freezing temperature θ_{if} (°C):

$$Q_1 = m\, C_p\, (\theta_i - \theta_{if}) \quad [2.3]$$

with m (kg) being mass and C_p (J kg^{-1}K^{-1}) the specific heat of the product. The energy used during crystallisation is:

$$Q_2 = m_w\, L \quad [2.4]$$

with m_w (kg) being the mass of freezable water of the product and L (J kg^{-1}) the latent heat of melting or freezing. Freezable water corresponds to the total water in the product from which strongly adsorbed water is subtracted. After crystallization, the frozen product is cooled to the final freezing temperature θ_{ff} (°C):

$$Q_3 = m\, C_{pf}\, (\theta_{if} - \theta_{ff}) \quad [2.5]$$

with C_{pf} (J kg^{-1} °C^{-1}) being the specific heat of the frozen product. Q_T is obtained by adding all these energy values:

$$Q_T = Q_1 + Q_2 + Q_3 \quad [2.6]$$

Generally, foods containing a small amount of water require a lower cooling capacity because the specific heat of the non-aqueous components is lower than that of water, and the quantity of freezable water is less.

At a fixed cooling capacity, it is important to rapidly cool the unfrozen product (first phase, quantity of energy Q_1) to prevent the growth of microorganisms. The cooling capacity of a food (temperature θ, mass m, area A and specific heat C_p) in a cooling chamber (temperature θ_c) can beexpressed according to Fourier's law [2.1]:

$$\dot{Q} = \frac{dQ}{dt} = A\ h_{BL}\ (\theta - \theta_c) \quad [2.7]$$

Where h_{BL} is the heat transfer coefficient of the boundary layer around the product. At each time interval dt, a quantity of elementary heat dQ is extracted, so that the product cools by dθ:

$$dQ = m\, C_p\, d\theta \quad [2.8]$$

By combining [2.7] and [2.8], the cooling rate $\frac{d\theta}{dt}$ is obtained, expressed as:

$$\frac{d\theta}{dt} = \frac{A}{m} \frac{1}{C_p} h_{BL} \ (\theta - \theta_c) \qquad [2.9]$$

Consequently, the cooling rate can be increased by:

– reducing the value of θ_c, which increases the value of the term $(\theta - \theta_c)$;

– dividing the food into portions, which increases the value of the term $\frac{A}{m}$;

– reducing the boundary layers (turbulent flow of the cooling medium) so as to increase the value of h_{BL}.

In practice, portioning the food and increasing the velocity of the cooling medium are the most effective methods.

2.1.3.2. *Freezing rate and its effect on food*

The freezing rate is a process control parameter as it determines the size of crystals and their distribution in the food (Figure 2.5). The growth rate of ice crystals depends on the heat extraction rate, that is the temperature difference between the crystal and its surrounding medium.

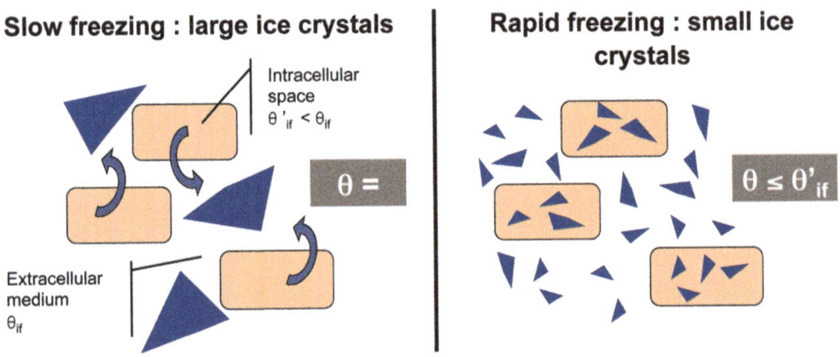

Figure 2.5. *Freezing rate and influence on product structure. θ_{if} and θ'_{if} represent the initial freezingtemperature of the extracellular medium and intracellular space, respectively*

If freezing is slow (< 1°C min^{-1}), extracellular crystallization, which increases the local concentration of solutes, causes a gradual dehydration of cells by osmosis. In this case, freeze concentration that occurs outside the cells results in an osmotic imbalance (the ionic force and osmotic pressure of the external medium are higher than inside the cells) leading to a migration of water to the outside. This transfer, which may have irreversible effects if it exceeds a certain level, largely explains the drop in turgidity, the separation of tissue and the exudation of water that occurs during the thawing of food (plasmolysis); it is the main cause of softening of plant tissue. Large ice crystals form and expand the extracellular spaces, while the plasmolysed cells decrease considerably in volume. These large crystals cause mechanical compression, which tends to crush the cells.

In contrast, when freezing is rapid, crystallization occurs simultaneously in both intra- and extracellular spaces. There is little water movement, and a large number of small crystals form. As a result, cell and tissue structure are much less affected. Furthermore, there are considerably fewer changes in texture due to the osmotic migration of water from cells compared to slow freezing. During thawing, products that are rapidly frozen exude much less than products that are slowly frozen.

The freezing period (corresponding to the thermal arrest time, i.e. the formation of ice) is therefore a critical phase for the final quality of the product, which is better the shorter the thermal arrest time or freezing plateau. It can be predicted using Planck's equation (Figure 2.6). For simplification purposes, it is assumed that the product is in the form of plates that are $2e$ in thickness, thin compared to the other dimensions (length and width), which can therefore be assimilated to infinite plates of area A. The initial temperature of the plate is assumed to be uniform and equal to the initial freezing temperature of the product (θ_{if}). The cooling phase corresponding to the loss of sensible heat from the product is not of interest here. The product characteristics (density ρ, latent heat of freezing L, thermal conductivity in the frozen state λ_f) as well as the freezing conditions outside the product (heat transfer coefficient h_{BL} of the boundary layer around the plate and the chamber temperature θ_c) are assumed to be known.

Figure 2.6 schematically shows the plate during freezing. The length of the plateau, where the temperature remains constant, is the time required for the freezing front to go from x = 0 at time 0 to x = e (half the thickness) at time t. This is why it is sufficient to consider a half plate.

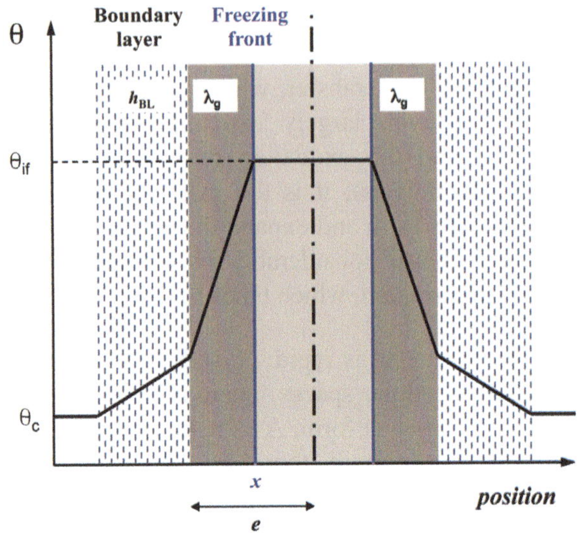

Figure 2.6. *Calculation of freezing time for an infinite plate*

The energy dQ released at the freezing front at a time dt moves the freezing front ahead by dx. This results in the freezing of a mass fraction dm such that:

$$dQ = dm\, L = \rho\, A\, dx\, L \qquad [2.10]$$

Moreover, the power exchanged can be expressed as in [2.7] according to Fourier's law:

$$\dot{Q} = \frac{dQ}{dt} = A\, h_g\, (\theta_{if} - \theta_c)$$

With h_g the overall heat transfer coefficient expressed as:

$$\frac{1}{h_g} = \frac{1}{h_{BL}} + \frac{x}{\lambda_f}$$

Combining these three equations, and integration over time, can give the freezing time t, known as Planck's formula:

$$t = \frac{\rho L}{\theta_{if} - \theta_c}\left(\frac{e}{h_{BL}} + \frac{e^2}{2\lambda_f}\right) = \frac{\rho L}{\theta_{if} - \theta_c}\, e\left(\frac{1}{h_{BL}} + \frac{e}{2\lambda_f}\right) \qquad [2.11]$$

The relative importance of the terms $\frac{1}{h_{BL}}$ and $\frac{e}{2\lambda_f}$ is variable depending on the product characteristics and the freezing mode used. In the case of static freezing, h_{BL} is small (generally in the range 10 W m^{-2} K^{-1}) and since the term $\frac{1}{h_{BL}}$ is large, the term $\frac{e}{2\lambda_f}$ can be neglected. In this case, the limiting factor is the heat exchange on the product surface, and the freezing time is therefore proportional to *e*. Conversely, if h_{BL} is large (freezing by immersion in brine or to a lesser extent forced air freezing), the limiting factor is the heat exchange within the product, and therefore the freezing time is proportional to e^2. This is also the case if thickness *e* is large, or the thermal conductivity of the material λ_f is very low (fats in particular), and it is not advisable in this case in trying to improve the heat exchange on the surface.

However, in most cases, both terms require consideration, and the freezing time is proportional to a second order polynomial function of *e* without a constant term. Thus, in order to reduce the length of the freezing plateau, it is possible to:

– increase the temperature difference (θ_{if} - θ_c);

– decrease the thickness (2*e*) of the product;

– increase the heat transfer coefficient h_{BL} of the boundary layer around the product.

However, it is also influenced by intrinsic factors of the product (ρ, λ_f, L). The approach presented here for a flat plate may be generally applied to other geometric models, and in the same way it is shown that the length of the freezing plateau is equal to:

$$t = \frac{\rho L}{\theta_{if} - \theta_c}\,\frac{R}{2}\left(\frac{1}{h_{BL}} + \frac{R}{2\lambda_f}\right) \qquad [2.12]$$

in the case of an infinite cylinder of radius R, and:

$$t = \frac{\rho L}{\theta_{if} - \theta_c} \frac{R}{3} \left(\frac{1}{h_{BL}} + \frac{R}{2 \lambda_f} \right) \quad [2.13]$$

in the case of a sphere of radius *R*.

For packaged products, the thermal resistance of the packaging as well as any unfilled volumes inside the packaging must also be taken into account.

2.1.3.3. *Changes in food during freezing and storage*

Changes in residual liquid spaces

One of the consequences of freezing is an increase in the solute concentration of the liquid portion of food. Therefore, the main physico-chemical changes resulting from freeze concentration include:

– an increase in salt concentration and a drop in pH due to an increase in the ionic force;

– changes in the redox potential, osmotic pressure, surface tension, vapor pressure or freezing point.

These changes, associated with the elimination of some liquid water, often have a negative impact on the quality of the food: a decrease in the solubility of protein (e.g. milk caseins and egg yolk lipoproteins) and the stability of elements dispersed in a colloidal or globular state (aggregation of fat globules in frozen dairy products). The aggregation of fish muscle proteins, and to a lesser extent meat proteins, induced by its prolonged storage at insufficiently low temperatures, causes increased exudation upon thawing and a hardening of the texture. The drop in water retention capacity and the hardening of the texture of the fish muscle after frozen storage is partly attributed to the aggregation of actomyosin. Hydrated carbohydrate gels, e.g. pectin, may also undergo destabilization as a result of frozen storage (flocculation of suspended material in certain fruit juices). Increasing the concentration of hydrophilic colloids (starch, pectin, gelatine) can considerably increase the viscosity of the product. Salts and sugars can also be precipitated (lactose in ice-cream). The result of all these changes is a destabilization during thawing.

It is possible to limit these consequences by rapid freezing and storing products at very low temperatures. The advantage of rapid freezing is partly due to the fact that some components are immersed in hypertonic solutions for a shorter time. The benefit of ultra-low temperature storage is that the products are almost completely solidified. However, storage temperatures below –20°C lead to high energy costs.

Changes in food during frozen storage

Frozen foods are not inert. Their quality gradually decreases during storage as a result of chemical and physical changes. The maximum storage periods that can be achieved for various foods, while maintaining quality at a satisfactory level (under suitable packaging conditions) depend largely on the temperature; they vary broadly on a log-linear scale depending on storage temperature, from less than one month at –7°C to almost two years at –30°C. In principle, the storage temperature should be chosen for each product based on its characteristics and expected shelf life (transit time, restocking time, etc.). Thus, a temperature of –30°C is sometimes necessary to store certain fish between 6 and 8 months. The same applies for ice cream in order to prevent the formation of large ice crystals during storage. These very low temperatures require a high level of energy, and are difficult to maintain during transport and at points of sale. In practice, most foods are stored at –18°C, which ensures an average shelf life of about one year.

Despite suitable freezing and storage conditions, a product cannot be preserved indefinitely, and its quality deteriorates over time due to a number of processes such as the development of a rancid flavor, discoloration, the loss of nutritional value, etc. Some of the deterioration reactions that occur during storage are catalysed by enzymes that can remain active even at very low temperatures. This is the case with the enzymatic browning of fruit and vegetables frozen in a raw state. The addition of ascorbic acid, sugar or sulphur dioxide as well as blanching before freezing helps to prevent browning during both freezing and storage.

The most notable non-enzymatic reactions that occur in frozen foods include the oxidation of lipids (oily fish, peas), vitamin C (strawberries), carotenoid pigments and flavor components. The degradation of anthocyanic pigments (strawberries) or chlorophyll (spinach, green beans, Brussels sprouts) is also a limiting factor to the shelf life.

Effect of storage temperature control on food quality

During cold storage, the frozen product undergoes temperature fluctuations that also contribute to a loss in quality. Despite being stored in a freezer at a very low temperature, refrigeration is an "all or nothing" mechanism, and the product is subjected to refrigeration cycles of varying amplitude. In each cycle, an increase in temperature results in the thawing of a certain amount of water in the product. This water refreezes during the following cycle when the compressor restarts.

These repeated cycles of water demobilization/mobilization cause a loss of small ice crystals in favor of large ice crystals (Oswald ripening or migratory recrystallization) resulting in a considerable decline in the quality of the frozen product over time. The gradual recrystallization of ice crystals, together with an increase in crystal size, adversely affects food texture (ice cream with a "sandy" texture) and appearance (frozen chicken with a glassy appearance). This can be avoided by producing the food in a way that its glass transition temperature is higher than the maximum storage temperature.

Temperature fluctuations also contribute to increasing the risk of surface dehydration, by removing the water from the product to the walls or cooling coils of the freezer. This dehydration causes weight loss (up to 3–5%) and promotes lipid oxidation (fish), hardening (fish, meat) and staining (brown or green spots on poultry, black spots on beef) referred to as "freezer burn".

Such dehydration and oxidation phenomena justify the need for waterproof, air-tight and light resistant packaging as well as vacuum or shrink packaging. Glazing, achieved by spraying water on the frozen product, also offers good protection. Other types of packaging are increasingly being used for frozen products: heat resistant packaging that allows the product to be thawed in boiling water or microwaves, foil plates for frozen ready-to-eat meals, etc.

2.1.3.4. *Thawing*

Thawing is slower than freezing because it involves the formation of an aqueous liquid layer on the surface and then on the outer layers of the product. Using the same approach as previously outlined, the thawing time can be calculated based on Planck's formula:

$$t = \frac{\rho \, L}{\theta_c - \theta_{it}} \, e \left(\frac{1}{h_{BL}} + \frac{e}{2\,\lambda} \right) \qquad [2.14]$$

with, in this case, θ_c (chamber temperature) greater than θ_{it} (initial thawing temperature), and λ the thermal conductivity of the unfrozen product. Since the latter is four times lower than that of the frozen product, most of the heat required for melting is only transferred slowly across the outer layer (aqueous) compared to freezing.

Maintaining the product at a temperature slightly below 0°C for extended periods of time has adverse effects. The product is exposed to relatively high concentrations of solutes, ice crystals become larger and the development of psychrophilic microorganisms is facilitated by an exudate rich in nutrients.

It is therefore best to thaw quickly. This is not a problem with vegetables, which can be dipped in boiling water, or with small pieces of meat of fish, which can be cooked without thawing beforehand. It is, however, more difficult in the case of fruit or large pieces of food (joints of meat or whole fish). Fruit is generally thawed at room temperature; meat and fish are best thawed in a refrigerator ($\theta < 10$°C) or cold water bath in order to reduce exudation and prevent the rapid growth of microorganisms.

Thawing by dielectric heating would be very advantageous if the inhomogeneity in the structure of most foods did not leave some parts frozen and other liquid parts overheated. In the case of fish, it saves a lot of time. Microwaves can also be used, but they often result in local variations in the thawing rate due to the heterogeneity of the product and the absorption of energy by the thawed parts. Therefore, low power is usually applied for limited time periods to allow the heat generated in the mass of the product to diffuse into cold and still frozen areas.

2.2. Concentration by evaporation

Concentration by evaporation involves placing a liquid in temperature and pressure conditions that allow the vaporization of the solvent. Using this method, non-volatile elements in the treated product can thus be concentrated. In the food industry, it is mainly used to remove water from solutions, emulsions and/or colloidal solutions.

A key aspect of this technique is the energy cost involved, as the removal of water in this case is obtained by phase transition (liquid–vapor), in contrast to separation techniques. The concentration at atmospheric pressure of 1 kg of a 10% sucrose solutionat 20°C to a 20% sucrose solution (removal of 0.5 kg of water) requires a total of 1,439 kJ. This energy can be broken down as follows: sensible heat to increase the temperature from 20 to 100°C (311 kJ) and latent heat to vaporise 0.5 kg of water at 100°C (1,128 kJ). Loncin [LON 61] noted that the energy difference between initial and final systems brought back to 20°C is only 0.5 kJ, which corresponds to the isothermal compression of sucrose molecules. The yield from concentration by evaporation is therefore very low without any energy recovery. Furthermore, most of the efforts made at a technical level were aimed at improving this yield.

In addition, liquids processed in the food sector are often heat sensitive. To reduce the biochemical spoilage of product components, concentration by evaporation is usually carried out under partial vacuum to reduce the processing temperature to a range of 45 to 80°C (Table 2.2). While the qualitative gain of this practice is obvious, the energy gain is in fact low. The spoilage of components, based on a time-temperature ratio, can also be reduced by lowering the residence time in the facility.

Firstly, the principle of vacuum evaporation is explained. Thereafter, the different techniques to reduce energy consumption are outlined.

2.2.1. *Single stage evaporation*

2.2.1.1. *Principle*

Single stage evaporation involves placing a liquid, which has been brought to boiling point, into a vacuum chamber (evaporator; Figure 2.7). The vacuum corresponds to the saturation vapor pressure at the boiling point of the product, e.g. at 70°C, Table 2.2 shows that the pressure in the evaporator is 31,156 Pa.

θ	P^0_v	H_v	L_v	ρ_v	θ	P^0_v	H_v	L_v	ρ_v
(°C)	(Pa)	(kJ kg^{-1})	(kJ kg^{-1})	(kgm^{-3})	(°C)	(Pa)	(kJkg^{-1})	(kJ kg^{-1})	(kg m^{-3})
0	610.8	2,500.4	2,500.4	0.005	**72**	33,960	2,629.7	2,328.7	0.215
20	2,337	2,537.2	2,453.5	0.017	**74**	36,961	2,633.1	2,323.7	0.232
30	4,241	2,555.6	2,430	0.030	**76**	40,188	2,636.4	2,318.6	0.251
36	5,940	2,566.5	2,415.8	0.042	**78**	43,649	2,639.8	2,313.2	0.272
38	6,624	2,570.3	2,411.2	0.046	**80**	47,356	2,643.1	2,308.2	0.293
40	7,375	2,573.6	2,406.2	0.051	**82**	51,328	2,646.5	2,303.2	0.316
42	8,198	2,577.4	2,401.5	0.056	**84**	55,574	2,649.4	2,297.7	0.341
44	9,100	2,580.7	2,396.5	0.062	**86**	60,105	2,652.8	2,292.7	0.367
46	10,085	2,584.1	2,391.5	0.069	**88**	64,949	2,656.1	2,287.6	0.394
48	11,162	2,587.9	2,386.9	0.076	**90**	70,108	2,659	2,282.2	0.424
50	12,335	2,591.6	2,382.3	0.083	**92**	75,609	2,662.4	2,277.2	0.455
52	13,613	2,595	2,377.3	0.091	**94**	81,464	2,665.7	2,272.2	0.488
54	15,002	2,598.3	2,372.7	0.100	**96**	87,691	2,668.7	2,266.7	0.522
56	16,509	2,602.1	2,368.1	0.109	**98**	94,301	2,672	2,261.7	0.559
58	18,146	2,605.4	2,363	0.119	**100**	101,325	2,674.9	2,256.3	0.598
60	19,917	2,609.2	2,358.4	0.130	**105**	120,800	2,682.5	2,242.5	0.705
62	21,389	2,612.6	2,353.4	0.142	**110**	143,270	2,690	2,229.1	0.826
64	23,909	2,615.9	2,348.4	0.154	**115**	169,060	2,697.6	2,215.2	0.965
66	26,145	2,619.3	2,343.4	0.168	**120**	198,540	2,704.7	2,201	1.122
68	28,557	2,622.6	2,338.3	0.183	**125**	232,080	2,711.8	2,187.2	1.299
70	31,156	2,626.4	2,333.7	0.198	**130**	270,110	2,718.5	2,172.5	1.497

Table 2.2. *Saturation vapor pressure P^0_v (Pa), enthalpy H_v (kJ kg^{-1}), latent heat of vaporization L_v (kJ kg^{-1}) and density ρ_v (kg m^{-3}) of vapor as a function of temperature [KES 86]*

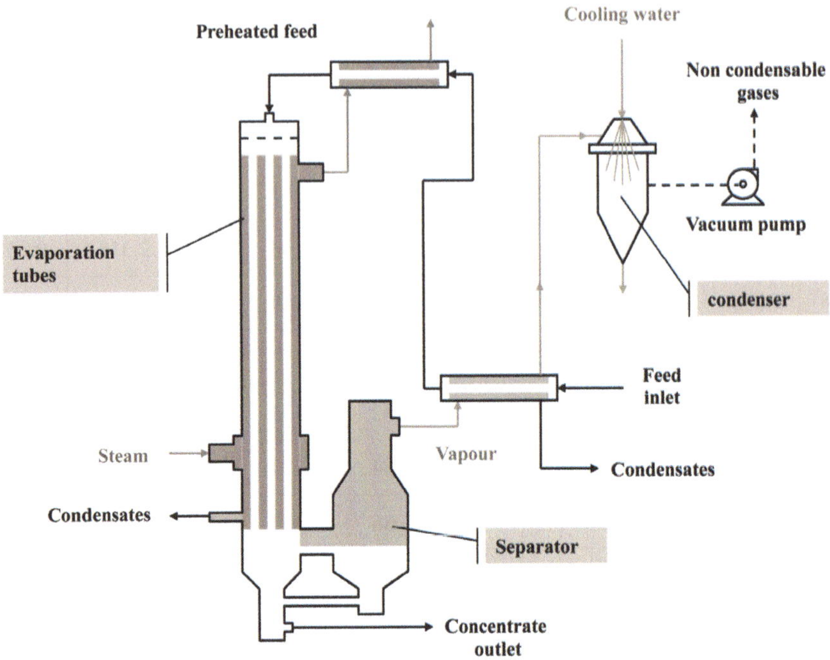

Figure 2.7. *Principle of a single effect evaporator*

In this case, any heat supplied to the product will result in the vaporization of some of the liquid. The evaporator therefore consists of a latent heat exchanger. In practice, the energy supplied to the heat exchanger (usually tube bundle) comes from primary vapor (also called steam) that is 5 to 10°C hotter than the product.

In the dairy sector, falling film evaporators are used, which means that the liquid, introduced at the top of the unit, flows down as a thin film (thickness in the range 0.1 to 1 mm) inside the tubes. Other types of evaporators exist, which vary in the circulation mode of the fluids (climbing film evaporators for sugar, etc.) or in the shape of the heating surfaces (e.g. plate surface).

The liquid–vapor mixture is separated in a vessel next to the evaporator. In this way, both the secondary vapor and the concentrated liquid are collected. Generally, the energy in the vapor is recovered, either to reheat the incoming product or to heat a second evaporator. This is the principle of multiple stage evaporation, which will be described later.

Figure 2.8 shows the energy balance in single effect concentration, which is divided between sensible heat to bring the product to boiling point (70°C in this case) and latent heat to evaporate part of the water. 1 kg of primary vapor at 80°C is needed to evaporate almost 1 kg of water at 70°C, corresponding to about 2,300 kJ of energy.

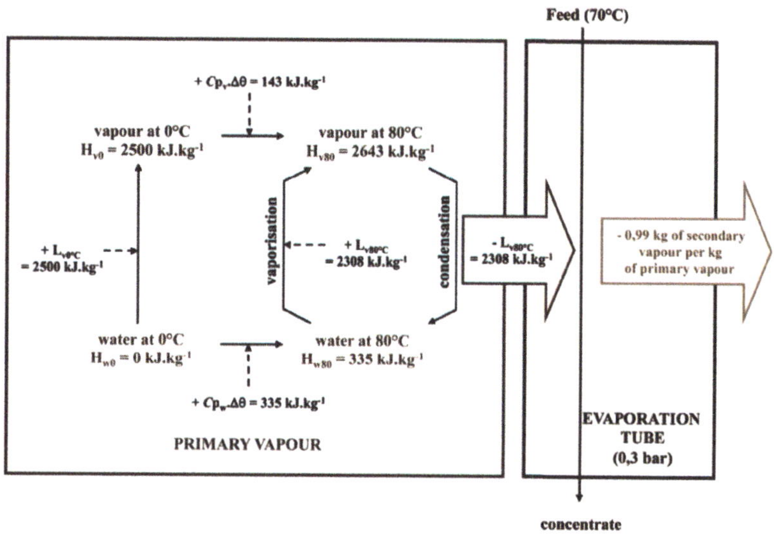

Figure 2.8. *Energy balance in single effect vacuum evaporation. H_w and C_{pw}: enthalpy and specific heat of liquid water; H_v and C_{pv}: enthalpy and specific heat of vapor; L_v: latent heat of vaporization*

2.2.1.2. *Vacuum generation*

In evaporators, vacuum is partly generated by the condensation of vapor. Condensation can be achieved by adding water to the vapor (direct condenser) or to the surface of a heat exchanger (indirect condenser) (Figure 2.9).

In order to limit the accumulation of non-condensable gases (mainly oxygen and nitrogen) from treated products, water of condensation or installation leaks, they must be extracted using a vacuum pump.

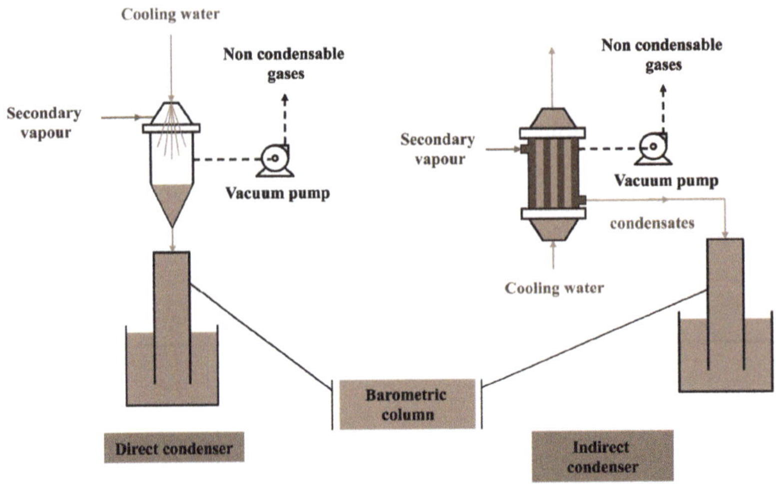

Figure 2.9. *Direct and indirect condensers*

It is possible to design these facilities based on an energy balance, according to the level of vacuum expected. On this basis, Jeantet *et al.* [JEA 01] show that the water mass flow rate $\dot{m}_w$ required to condense vapor at 50°C (mass flow rate $\dot{m}_{sv}$) and reach a vacuum of 0.012 MPa is equal to $19\,\dot{m}_{sv}$. As a result, this operation involves a considerable consumption of water. Water consumption can be reduced by using an indirect condenser, which is generally supplied by water in a closed circuit and cooled by an air cooler.

2.2.2. *Reduction in energy consumption*

The elimination of water is often an expensive operation in the processing of liquid foods. Several solutions exist to reduce the cost of concentration by vacuum evaporation.

2.2.2.1. *Multiple stage evaporation*

The multiple stage evaporator consists of a set of single-stage evaporation units arranged in series (Figure 2.10) whereby the liquid food being concentrated passes from one stage to the next. The first unit is heated with direct steam injection while the subsequent units are heated with the secondary vapor generated in the preceding stage. The last stage is connected to a condenser, creating a vacuum in the entire system. The energy required to eliminate water in an evaporator with n stage is therefore $\frac{2,300}{n}$ kJ kg^{-1} of water removed (not taking into account sensible heat).

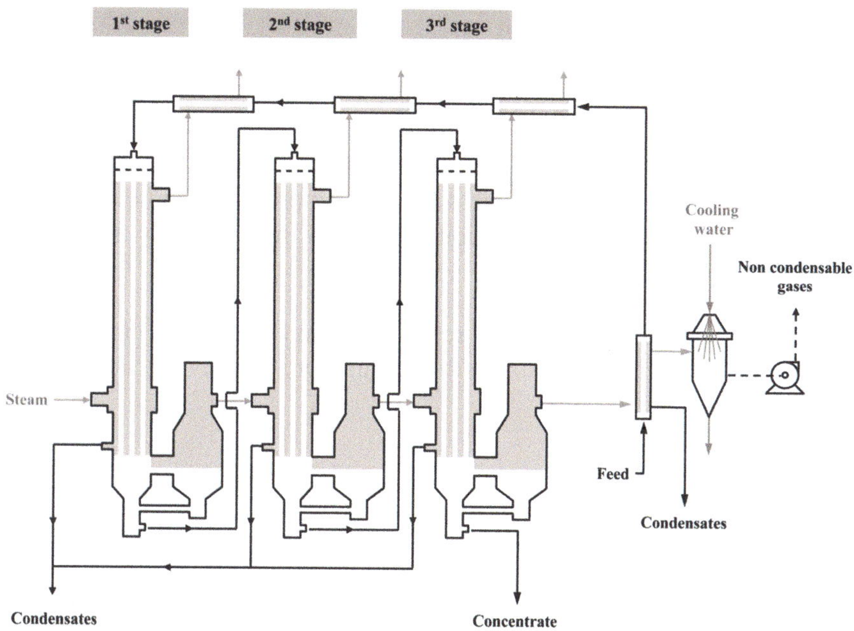

Figure 2.10. *Principle of a multiple stage evaporator*

Since a temperature difference is necessary between the steam and the product to be concentrated, the evaporation temperatureand therefore the pressure decrease from one unit to the next.

The limits to increasing the number of effects are:

– the maximum temperature that the product can withstand in the first effect, due to heat stability. In practice, this temperature is generally between 70 and 90°C for most foods;

– the temperature in the last effect, which is limited by the temperature of the water of condensation and/or the increase in viscosity due to the drop in temperature and the increase in dry matter of the concentrate. In practice, this temperature is higher than or equal to 40°C;

– the drop in the evaporation temperature from one stage to the next. This drop is generally greater than or equal to 5°C.

The compromise between reducing energy costs and investing in additional effects is an evaporation system with between three and six effects.

2.2.2.2. *Vapor recompression: thermal vapor recompression and mechanical vapor recompression*

The vapor generated can be recompressed in order to increase its enthalpy level, and then reused in the process. This recompression can be achieved by heat (thermal vapor recompression; TVR) or mechanically (mechanical vapor recompression; MVR).

The principle of thermal vapor recompression (Figure 2.11) is to increase the energy of the vapor by mixing it with high pressure live steam or motive steam.

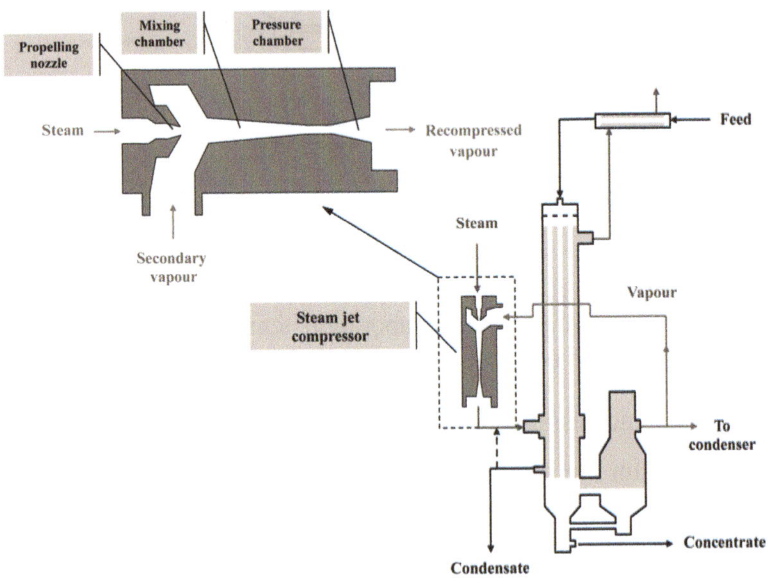

Figure 2.11. *Principle of a steam jet compressor*

By mixing 1 kg of steam (1 MPa and 180°C: H_v = 2,777 kJ kg^{-1}) with 1 kg of vapor (0.02 MPa and 60°C: H_v = 2,609 kJ kg^{-1}), the result is 2 kg of vapor (H_v = 2,694 kJ kg^{-1}). This recompressed vapor can be brought to saturation by adding the condensates (≅ 0.04 MPa and 78°C).

A thermocompressor is equivalent to one more stage in terms of energy cost reductions, but requires less investment and is smaller in size. Vapor generated in an evaporation stage can be returned to this latter after thermal compression (Figure 2.16) or even recycled in an upstream stage.

Mechanical vapor recompression (Figure 2.12) can also reduce energy costs. Vapor (P_1 ≅ 0.02 MPa, enthalpy H_1 = 2,609 kJ kg^{-1} and temperature θ_1 = 60°C) is mechanically compressed to a pressure of P_2 ≅ 0.03 MPa (enthalpy H_2 = 2,675 kJ kg^{-1}).

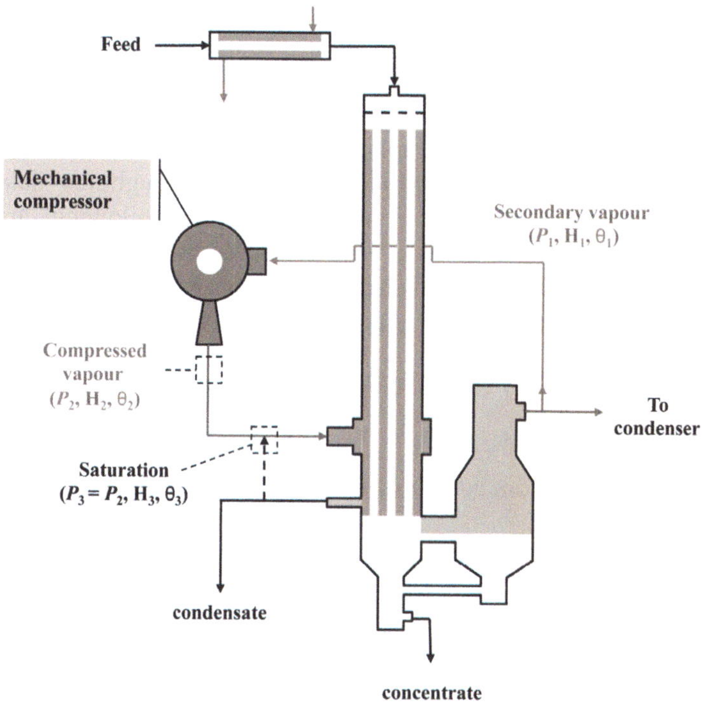

Figure 2.12. *Principle of mechanical vapor recompression*

Superheated vapor is brought to saturation by introducing condensates to facilitate its condensation and the release of latent energy. The resulting vapor is at $P_3 \cong 0.03$ MPa, $H_3 = 2{,}625$ kJ kg^{-1} and $\theta_3 = 69$°C.

Like thermal vapor recompression, mechanical vapor compression can be implemented on one stage or on several evaporation stages simultaneously. The energy consumption of mechanical vapor compression, taking into account the mechanical compressor performance, is $\frac{75}{0.9} \cong 85$ kJ kg^{-1} of water removed in a single stage and 30 kJ kg^{-1} of water removed in a triple stage. However, if the required energy is supplied by electricity, the primary energy consumption in this case is $30 \times 3 \cong 90$ kJ kg^{-1} of water removed. These energy costs can be compared with those of a three-stage or six-stage evaporator, which are 800 and 400 kJ kg^{-1} of water removed, respectively.

Today, the choice between thermal vapor recompression and mechanical vapor recompression is an economic compromise between energy saving and investment. This saving is equivalent to one or two additional effects. Table 2.3 illustrates this point by giving the energy consumption of concentration by vacuum evaporation based on the number of effects and whetheror not TVR and MVR are used. Although the data is not entirely comparable (depending on whether the following are taken into account: centrifugal pumps, vacuum pump, water of condensation, vapor, boiler efficiency, etc.), this table shows that increasing the number of effects and using TVR and MVR leads to a decreasein energy consumption.

To summarise, energy costs can be improved by:

– preheating solutions,

– recovering heat from condensates and concentrates,

– using vacuum,

– using a multiple effect system,

– using thermal vapor recompression or mechanical vapor recompression.

Number of stages	*TVR*	*MVR*	*Energy consumption (kWh t^{-1} of water removed)*
1	no	no	861
	no	no	731–772
2	no	no	476
	no	no	365–408
3	no	no	343
	no	no	248–289
	yes	no	140
	yes	no	164–185
	yes	no	278
	no	**yes**	55
	yes	**yes**	71
4	no	no	193–217
	yes	no	116–145
5	**yes**	no	101
	yes	no	111
	yes	no	93–116
6	**yes**	no	73–93
7	**yes**	no	70

Table 2.3. *Energy consumption of vacuum evaporators based on the number of effects and whether or not thermal vapor recompression (TVR) and/or mechanical vapor recompression (MVR) are/is used [KES 86, HAR 92, LEL 92, WES 94, MAF 96]*

2.3. Dehydration

Dehydrating food products improves stability by lowering a_w and reducing transport and storage costs Bimbenet and Loncin, [BIM 95]. This can be achieved by:

– evaporation at boiling temperature at atmospheric pressure or under partial vacuum (e.g. drying on heated rollers);

– sublimation of ice at partial pressures of water below the triple point (610.8 Pa), corresponding to the direct transition from a solid to a gaseous state (❶; Figure 2.13). This is the case for the freeze-drying of food products (coffee, mushrooms, etc.);

– evaporation (❷; Figure 2.13) under the combined action of a heat transfer from the hot air to the product and a water transfer from the product to the hot dry air (e.g. spray drying).

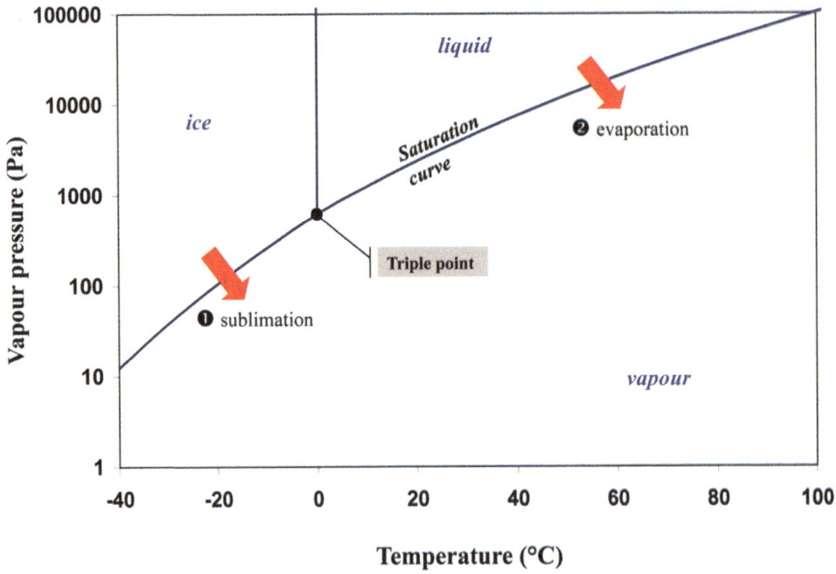

Figure 2.13. *Phase diagram of water*

2.3.1. *Roller drying*

2.3.1.1. *Principle*

Roller or drum drying involves transmitting a heat flux to the product, which has been brought to its boiling point, via a latent heat exchange surface. In the same way as for vacuum evaporation, the evaporation of water under these conditions is directly proportional to the energy input (latent heat of vaporization). In practice, this input is achieved by conduction via an exchange surface in contact with the product by vapor at a temperature of between 130 and 150°C. According to Fourier's law, heat transfer is proportional to the temperature difference between the heat transfer fluid (which usually remains at a constant temperature) and the boiling liquid at a given pressure.

2.3.1.2. *Energy*

Attempts have been made to model processes using drum dryers [VAS 84]. Since the heating surface is a set characteristic for a given device, the user can vary two factors to modulate the evaporation capacity: the temperature of the heating steam and the overall heat transfer coefficient (by reducing the thickness of the layer).

The specific energy consumption of roller drying is about 900 kWh t^{-1} of evaporated water. Pilot scale drum dryers in a partial vacuum chamber with mechanical vapor recompression can reduce energy consumption by up to 250 kWh t^{-1} of water removed.

Drum drying is therefore a valuable technique in terms of energy. In addition, it is possible to treat products at high initial concentrations. For example, a sodium caseinate with a dry matter of 400 g kg^{-1} dries easily on drums despite its high viscosity.

2.3.1.3. *Material*

The drum dryer consists of two drums positioned horizontally next to each other and heated internally by steam (Figure 2.14). The paste-like liquid is poured between the drums that rotate slowly in opposite directions. A film forms on the surface of the drums, which quickly dries and is removed by a scraper blade. The secondary vapor is extracted into a hood above the drums.

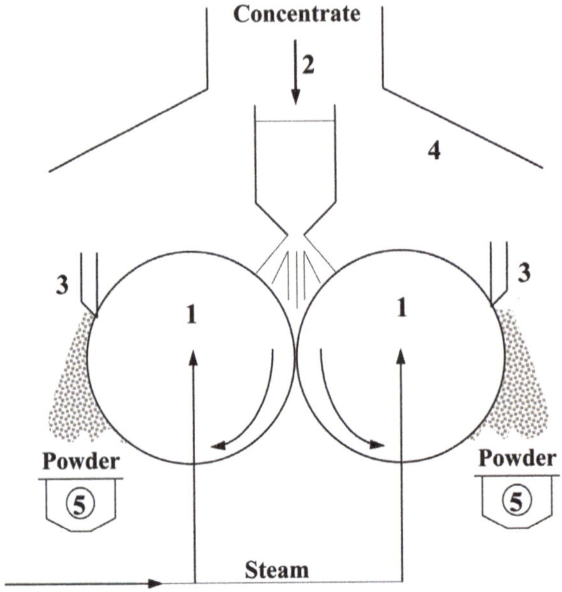

Figure 2.14. *Roller dryer. 1: Drum, 2: Feed, 3: Knife, 4: Vapor hood, 5 : Conveyor.*

This method is used in the starch industry, in the preparation of fruit and vegetable flakes (potato, cassava, etc.) and, to a lesser extent, in the dairy industry. Oxidation of the product is limited during drying, as a protective layer of secondary vapor covers the roller for most of its rotation. However, roller drying does not always meet quality requirements (solubility > 99%, dispersibility > 95%, low protein denaturation and reduced free fat content, etc.) due to the high heat treatment (temperature between 120 and 130°C and residence time between 2 and 20 s). In some cases, the physicochemical changes induced by heat are desired, e.g. powders for the biscuit and chocolate industry.

2.3.2. *Spray drying*

Spray drying involves placing wet particles in a sufficiently hot and dry stream of air (or another gas). Under these conditions, the difference in temperature and partial pressure of water between the particles and the air results in:

– a heat transfer from the air to the product due to the temperature difference;

– a water transfer from the product to the air due to the difference in the partial pressure of water between the air and the surface of the product (Figure 2.17).

This is the most commonly used drying method across all sectors of the food industry (charcuterie, fish products, fodder, cereals and vegetable products, fruit, milk, eggs, blood, etc.).

2.3.2.1. *Thermodynamics of humid air*

Air containing water in the form of vapor can be defined thermodynamically according to three parameters: absolute (AH) and relative humidity (RH), temperature (θ) and enthalpy (H).

Absolute and relative humidity

Absolute humidity is defined as:

$$AH = \frac{m_w}{m_a} \qquad [2.15]$$

where m_w is the quantity of water and m_a is the quantity of dry air. AH is usually expressed in kg of water per kg of dry air. Where water vapor behaves like an ideal gas, the Clapeyron equation can be applied:

$$PV = nRT \qquad [2.16]$$

where P is the pressure of the gas (Pa), V is the volume of the gas (m^3), n is the number of moles (mol), R is the perfect gas constant (J mol^{-1} K^{-1}) and T is the absolute temperature (K). Equation [2.16] can be rewritten when V and T are constant:

$$n = \frac{m}{M} = \frac{PV}{RT} = k\,P$$

m being the mass of the gas and M is its molar mass, and k being equal to $\frac{V}{RT}$. It can be deduced that:

$$m = k\,P\,M \quad [2.17]$$

If P_v is the partial pressure of water vapor (M = 18 g mol^{-1}) for a total pressure of the mixture P_t, the partial pressure of air (M = 28.9 g mol^{-1}) is $P_t - P_v$. Combining equations [2.15] and [2.17] leads to an expression of AH as a function of P_v and P_t:

$$AH = \frac{18\,P_v}{28.9(P_t - P_v)} \quad [2.18]$$

Relative humidity H_R is defined as:

$$RH = \frac{P_v}{P_v^0} \quad [2.19]$$

with P_v^0 being the saturated vapor pressure (in Pa) at air temperature θ. This value is often expressed as a percentage (%) of saturation. By combining [2.18] and [2.19], we obtain:

$$AH = \frac{18\,RH\,P_v^0}{28.9(P_t - RH\,P_v^0)} \quad [2.20]$$

Temperature

Air temperature θ can also be referred to as dry-bulb temperature, as opposed to its wet-bulb temperature θ_{wb}, which can be defined as the temperature at equilibrium of humid air cooled to saturation (formation of first drops of liquid water) by increasing AH. In practice, it is the temperature given by a wet-bulb thermometer (where the thermometer bulb is covered in a porous material impregnated with water; humid air moves around it at a sufficient speed to reach an equilibrium defining θ_{wb}). θ_d is the temperature to which humid air must be cooled (at constant AH) to reach saturation i.e. dew-point temperature. θ_d and θ_{wb} are less than or equal to θ.

Enthalpy

The thermodynamic properties of humid air at atmospheric pressure are determined by considering 1 kg of dry air containing AH kg of water. The system [dry air and water] is at a temperature θ. The enthalpy of humid air H is the sum of the enthalpy of air H_a and the enthalpy of the water vapor H_v, starting from 0°C.

The enthalpy of the air H_a corresponds to the energy required to bring 1 kg of dry air at 0°C to θ:

$$H_a = C_a.\theta \qquad [2.21]$$

where C_a is the specific heat of the air, equal to 1.01 kJ kg^{-1}°K^{-1}. The enthalpy of water vapor H_v corresponds to the energy required to evaporate AH kg of water at 0°C, and heat AH kg of vapor from 0°C to θ:

$$H_v = AH\left(L_0 + C_v\,\theta\right) \qquad [2.22]$$

Where C_v and L_0 are, respectively, the specific heat of water vapor and the latent heat of vaporization at 0°C, equal to 1.93 kJ kg^{-1}°K^{-1} and 2,500.4 kJ kg^{-1}, respectively.

The enthalpy H_m of the system is therefore equal to:

$$H_m = C_a\ \theta + H_A\ (L_0 + C_v\ \theta) = 1.01\ \theta + H_A\,(2{,}500.4 + 1.93\,\theta) \qquad [2.23]$$

2.3.2.2. *Enthalpy diagram of humid air*

The enthalpy of humid air, defined in [2.23], may be rewritten as:

$$Y = H - 2{,}500.4\ AH = 1.01\ \ \theta + 1.93\ \theta\ AH \qquad [2.24]$$

From the first term of [2.24], it is possible to plot curves at constant enthalpy H (isenthalps) as a function of AH. For a fixed enthalpy H, these functions have a slope of –2,500.4 with an intercept (AH = 0) of H. Therefore, the isenthalps are parallel lines.

Isotherms (curves at constant θ) are obtained from the second term of [2.24] as a function of AH. The isotherms, like isenthalps, are straight lines. The slope of these lines, zero for θ = 0°C, increases with θ. Unlike isenthalps, the isotherms intersect the y-axis (AH = 0) at a value slightly greater than θ.

Furthermore, it is possible to plot θ as a function of AH for different values of RH, knowing the relationship between AH, P_v^0 and RH (equation [2.20]) and the relationship between P_v^0 and θ (Table 2.2). The enthalpy diagram of humid air (Figure 2.15), also known as the Mollier diagram, can be obtained by superimposing isenthalps, isotherms and curves at constant RH.

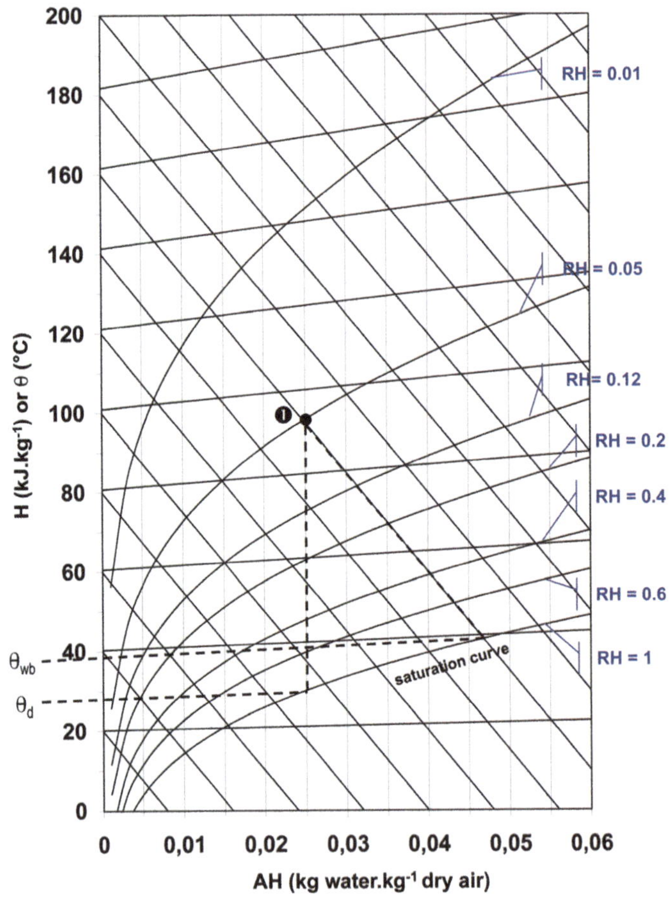

Figure 2.15. *Enthalpy diagram of humid air*

For example, the properties at point ➊ in Figure 2.15 are as follows:

– absolute humidity AH = 0.03 kg kg^{-1}: this value can be read on the x-axis below point ➊;

– enthalpy H = 160 kJ kg^{-1}, read directly on the isenthalp;

– relative humidity RH = 10%, read directly on the curve at constant RH;

– dry-bulb temperature θ = 80°C, read directly on the isotherm;

– wet-bulb temperature θ_{wb} = 39.2°C, at the intersection of the isenthalp H = 160 kJ kg^{-1} and the saturation curve RH = 100%;

– dew-point temperature θ_d = 31.6°C, read on the isotherm at the intersection between the saturation curve RH = 100% and a vertical line below point ➊.

The area under the saturation curve corresponds to mist region, a mixture of saturated air and water droplets or ice particles depending on the temperature.

2.3.2.3. *Properties of the enthalpy diagram*

Drying

Figure 2.16 describes an example of an air cycle during the drying of a liquid by entrainment. Air at 20°C with absolute humidity AH = 0.01 kg kg^{-1} (point ➊; H = 46 kJ kg^{-1}; RH = 70%) is heated to θ = 180°C (point ➋). Its enthalpy H_m is therefore 210 kJ kg^{-1} (AH = 0.01 kg kg^{-1}; RH = 0.16%). Upon contact with the liquid, there is a heat transfer from the air to the product and vice versa. The enthalpy of dry air decreases and that of water vapor increases. In first approximation, it is possible to assume that the overall enthalpy of the system remains close to constant (assumed to be adiabatic drying), which implies that the bond between water and other components remains weak and that it can be dried at a similar energy cost to latent heat of evaporation. In practice, this ideal case does not apply when drying many food products, leading to overheating of the air from 20 to 30°C for the same water removal.

If the temperature of the outlet air is 86°C, the absolute humidity of this air is 0.047 kg kg^{-1} (point ➌; RH = 12%). The quantity of evaporated water per kg of dry air is therefore 47 – 10 = 37 g. The removal of 1 kg of water in this case requires $\frac{1}{0.037}$ = 27 kg of dry air, or an energy cost of 27×(210 – 46) = 4,428 kJ kg^{-1}.

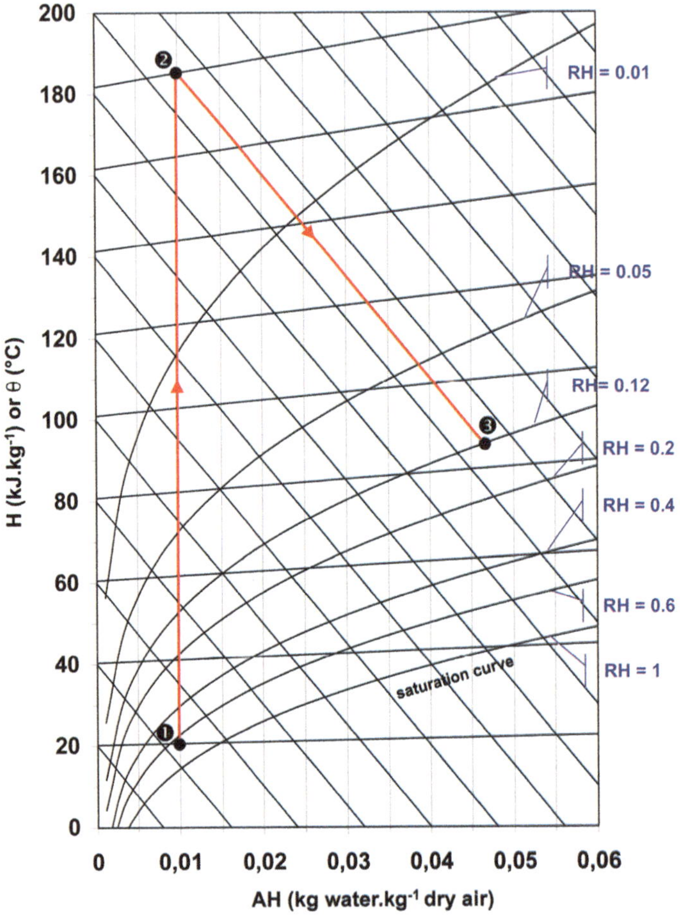

Figure 2.16. *Application of the enthalpy diagram of humid air for spraydrying*

Mixture of airs

Using the enthalpy diagram of humid air, it is possible to determine the resulting properties of a mixture of two predefined types of air (Table 2.4).

	Mass (kg dry air)	*Enthalpy ($kJ\ kg^{-1}$)*	*Absolute humidity (kg water kg^{-1} dry air)*
Air 1	m_1	H_1	AH_1
Air 2	m_2	H_2	AH_2
Mixture M	$m_1 + m_2$	H_m	AH_m

Table 2.4. *Properties of air 1 and air 2 and their mixture M*

The absolute humidity H_{Am} and enthalpy H_m of the mixture M are such that:

$$\begin{cases} m_1\ AH_1 + m_2\ AH_2 = (m_1 + m_2)\ AH_m \\ m_1\ H_1 + m_2 H_2 = (m_1 + m_2)\ H_m \end{cases} \qquad [2.25]$$

which leads to:

$$\frac{m_1}{m_2} = \frac{AH_2 - AH_m}{AH_m - AH_1} = \frac{H_2 - H_m}{H_m - H_1} \qquad [2.26]$$

Geometrically, an equal ratio in equation [2.26] results in the vertical alignment of points 1, 2 and M in the enthalpy diagram: point M is exactly between points 1 and 2. The mixture of the two unsaturated airs is located in the "mist" section, given the concavity of the saturation curve.

2.3.2.4. *Drying kinetics*

Figure 2.17 represents a water droplet suspended in air: P_v is the partial pressure of water at θ, and P_v^0 is the saturated vapor pressure at θ_{wb}.

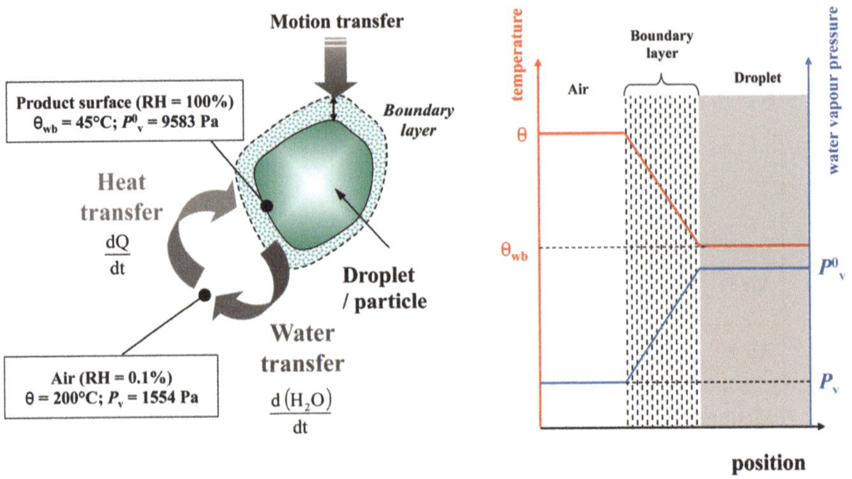

Figure 2.17. *Coupled heat and water transfer between a water droplet and ambient air*

Under stationary conditions, the transfer of water in air can be written as follows, according to Fick's law:

$$\frac{d\left(H_2O\right)}{dt} = A\ k\ \left(P_v^0 - P_v\right) \qquad [2.27]$$

where A is the exchange surface (m^2) and k is the water transfer coefficient from the droplet to the air (kg m^{-2} s^{-1} Pa^{-1}, or s m^{-1}). Since θ_v is less than θ, the heat transfer from air to the droplet can be expressed according to Fourrier's law under stationary conditions:

$$\frac{dQ}{dt} = A\ h\ \left(\theta - \theta_{wb}\right) \qquad [2.28]$$

Where h is the heat transfer coefficient between the droplet and the air (W m^{-2} °K^{-1}). If the process is adiabatic (constant enthalpy of the system), that is air is cooled by absorbing moisture, the following equation also applies:

$$dQ = L_v\ d\left(H_2O\right) \qquad [2.29]$$

where L_v is the latent heat of vaporization of water (J kg^{-1}). By combining [2.27], [2.28] and [2.29], we obtain:

$$P_v^0 - P_v = \frac{h}{L_v\, k}(\theta - \theta_{wb}) \quad [2.30]$$

h and k are highly dependent on air velocity, which determines the characteristics of heat and mass exchange from the peripheral boundary layer to the droplet (Figure 2.17), whereas $\frac{h}{L_v\, k}$ remains almost constant for air velocities ranging between 2 and 10 m s^{-1}, where its value is 64.7 SI units.

Drying kinetics can therefore be described, under stationary conditions, by equation [2.30]. Drying is the evaporation of surface water made available by the capillary rise of water inside the product to the surface. The rate of evaporation at the start of the drying process is constant as long as the average moisture is sufficient to remove water from the product surface. The partial pressure of water vapor of the product is therefore equal to P_v^0 and its temperature is equal to θ_{wb}, the wet-bulb temperature. At the end of the drying process, the rate of evaporation decreases since the partial pressure of water vapor of the product decreases towards P and its temperature increases towards θ, the dry-bulb temperature.

In the case of continuous drying, kinetics differ depending on whether the mode is co-current or counter-current (Figure 2.18).

Counter-current drying can achieve greater dehydration levels but is not recommended for heat-sensitive products since the outgoing product is heated to temperatures close to those of the incoming air. Moreover, the temperature difference between the product and its glass transition temperature is greatest in this case, which affects biochemical and physicochemical properties (see Volume 1 [JEA 16a], Chapters 1 and 7). Co-current drying, which is less harmful in terms of product quality, is also usually faster.

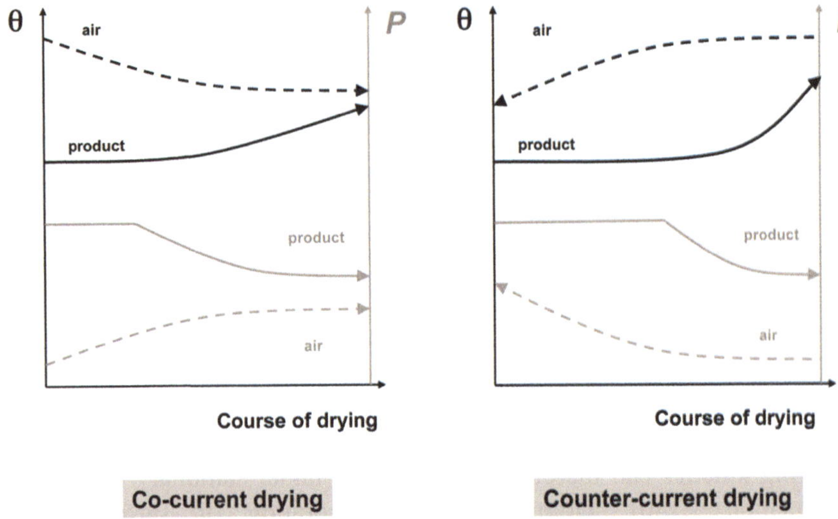

Figure 2.18. *Co-current or counter-current drying kinetics*

2.3.2.5. *Equipment*

There are many different types of dryers and drying processes depending on product properties, equipment and types of drying:

– *state of dispersion of the product*: a finely dispersed product dries faster (e.g. aerosol: spray drying) by increasing the exchange surface;

– *dynamic of the product (rotary dryer) and/or heating fluid*: the aim is to increase the $\dfrac{h}{L_v\, k}$ coefficient by reducing the limiting factors, which results in faster, more uniform drying. However, an overly turbulent state can cause excessive fragmentation of the product. In the case of air heating, the product is often travelling in an air flow that is itself in motion (tunnel dryer, fluidised bed or spray drying, etc.);

– *continuous and batch drying*: continuous drying is necessary for large-scale production. It requires more expensive equipment, but generally reduces the operating cost (labour). In particular, it helps to maintain the drying characteristics more constant in terms of the time and position in the dryer: air and product temperature, final water content of the product and energy efficiency of the dryer;

– *partial recycling of air*: this improves energy efficiency, ensures more uniform drying and maintains constant drying conditions;

– *air heating mode*: direct heating by mixing with flue gases (gas) or indirect heating by using heater batteries (electrical or supplied by a hot medium – steam, gas, oil). The choice of heating mode also depends on the physical (solid varying in size, liquid varying in viscosity) and chemical properties (sensitivity to heat or oxidation) of the product.

The main types of entrainment dryers are described below:

– *tray dryer*: food is spread in a thin layer on trays and heated, usually by hot air circulating between the trays (2 to 8 m s^{-1}). The air is generally recycled, especially at the end of the drying process, and its flow direction can be reversed occasionally to obtain a more uniform drying. This type of multipurpose batch dryer is generally inexpensive.

– *tunnel dryer*: this is essentially a tray dryer that operates on a continuous basis. The product passes through a tunnel in which hot air circulates. It usually has two sections (the first where air and the product flow co-currently and the second where they flow counter-currently), which combines the advantages of rapid initial drying and a low final water content. The product (and air) has a constant temperature and humidity at each position in the tunnel. These dryers are primarily used for drying fruit and vegetables. There are different types of tunnel dryers:

- *rotary dryer*: the product flows into a slightly tilted rotary cylinder (Figure 2.19). The air flow rate must be limited so as not to entrain product particles. This type of dryer allows rapid and uniform drying and is used for sugar, cocoa beans or meat,

- *belt dryer*: the product is evenly distributed on a perforated metal belt. Hot air flowseither parallel to the product or, more commonly, perpendicularly to the perforated belt and through the product layer. Drying is therefore uniform, especially since the movement of the product is often increased by using several belts positioned underneath each other (Figure 2.20).

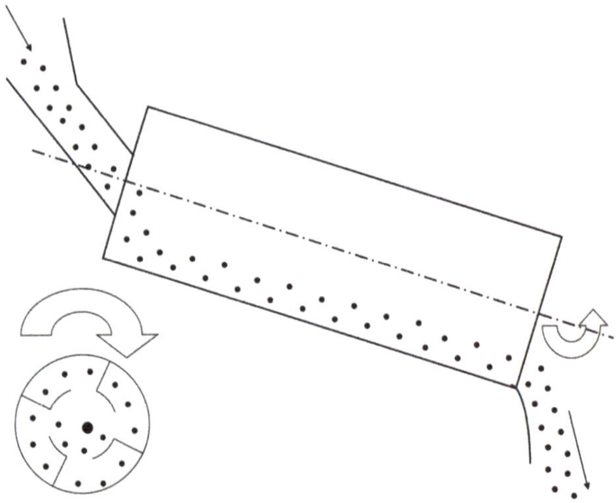

Figure 2.19. *Rotary dryer. Longitudinal and cross sections*

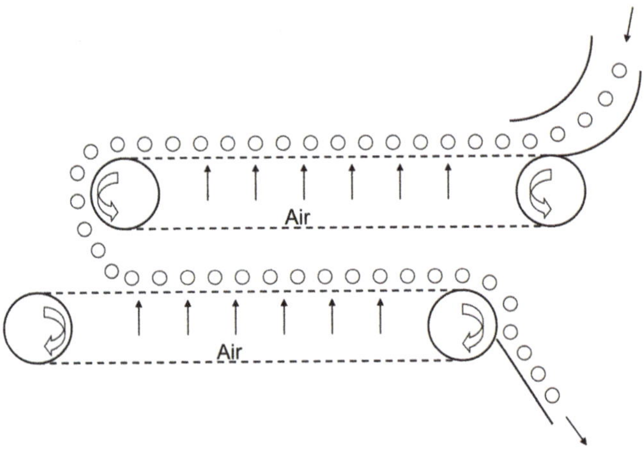

Figure 2.20. *Belt dryer*

–*fluidized bed dryers*: a layer of solid product, in the form of suitably sized particles, is placed on a porous plate or grid and subjected to an upward air flow that is fast enough to keep the product particles moving and suspended. Drying is generally continuous with the product layer moving from the inlet to the outlet. Residence time dispersion can however be problematic since some particles may be over or under heated. Drying is rapid due to the large air–product contact surface. The duration of drying decreases with particle size due to water diffusion; however, smaller particles can be entrained by the air flow, which makes it necessary to filter the outlet air by passing it through a cyclone separator.

This type of dryer is used for vegetables, cereals, coffee, sugar and milk powders. Some foods cannot be fluidised unless the water content is already relatively low:

– *spray drying (atomization)*: this involves spraying the product (liquid or suspension) into a stream of hot dry air. The air serves both as a heating medium and as a vector for removing moisture: it is hot and dry upon entering the drying chamber and cold and moist when it leaves. The drying rate is proportional to three factors:

- the exchange and evaporation surface, increased by spraying: 1 litre of sprayed liquid with droplets of 100 μm in diameter exhibits a surface of 60 m^2, whereas this is only approximately 5 dm^2 for an equivalent sphere of the same volume;

- the difference between the partial pressures of water in the particles and in the drying air: drying the air and/or raising its temperature can increase this value;

- the diffusion rate of water from inside the droplet to its surface, which depends on the diffusion coefficient and varies with the biochemical composition, water content and temperature of the product. This factor is essential for preserving the quality of the dried product since the presence of water on the particle surface can limit its temperature to the wet-bulb temperature for a long time.

A spray drying facility comprises a liquid and hot air circuit. Atmospheric air is drawn through filters, the type of which depends on local conditions and the nature of the product. The drying chamber is either cylindrical or cylindro-conical in shape, and can operate as a separator where most of the product is recovered at the base. Its size depends on the spray system: centrifugal atomiser, liquid pressure nozzle or two-fluid nozzle. If the dryer has a centrifugal atomiser, the chamber's height is less than or equal to its diameter. At the air outlet, a separator recovers the entrained fine particles: facilities are generally equipped with cyclones, air washers or bag filters.

Single-stage spray dryer is designed so that the residence time in the drying chamber is very short (20 to 60 seconds on average), and less than needed to reach the equilibrium between the moisture in the air and the product. The outlet temperature of the air should therefore be higher and the energy efficiency of the operation lower.

Two-stage spray drying involves limiting the extent of drying in the chamber in favor of that on an external fluidised bed, leading to a longer residence time (several minutes) and to outlet conditions closer to thermodynamic equilibrium. The product at the outlet of the spray drying unit should have a high enough moisture content to allow continuous flow and ensure good operating conditions. This causes a significant decrease in outlet air temperature and an increase in inlet air temperature. To obtain the required residual moisture, final drying takes place in an external fluidised bed or vibro-fluidiser where the air flow rates and processing temperatures are lower than in the drying chamber and therefore better suited to preserving the quality of the powder. One way of reducing costs and improving operating performance is to perform most of the drying in the fluidization phase rather than the atomization phase. The fact that the product sticks to the walls of the drying chamber is however a limiting factor. This contact is inevitable given the internal turbulence in the chamber.

This drawback can be eliminated in "three-stage" dryers (MSD: Multi Spray/Stage Drying). The design of the atomization phase and the positioning of the internal fluidised bed at the bottom of the atomization chamber help to limit contact between the moist product and the walls of the chamber (Figure 2.21).

Optimising the process and eliminating its initial limitations has remarkably improved performance and product quality:

– improved energy efficiency: significant decrease in outlet air temperature and increase in the outlet air relative humidity, resulting in a significant increase in inlet air temperature;

– reduced equipment size: in a given volume, the capacity is two to three times greater than in a conventional unit;

– substantially reduced atmospheric emissions: decrease in drying air flow rate and increase in moisture content of the product causing a reduction in loss by entrainment;

– improved product quality in terms of agglomeration, solubility, dispersibility, wettability, particle size, density, etc.

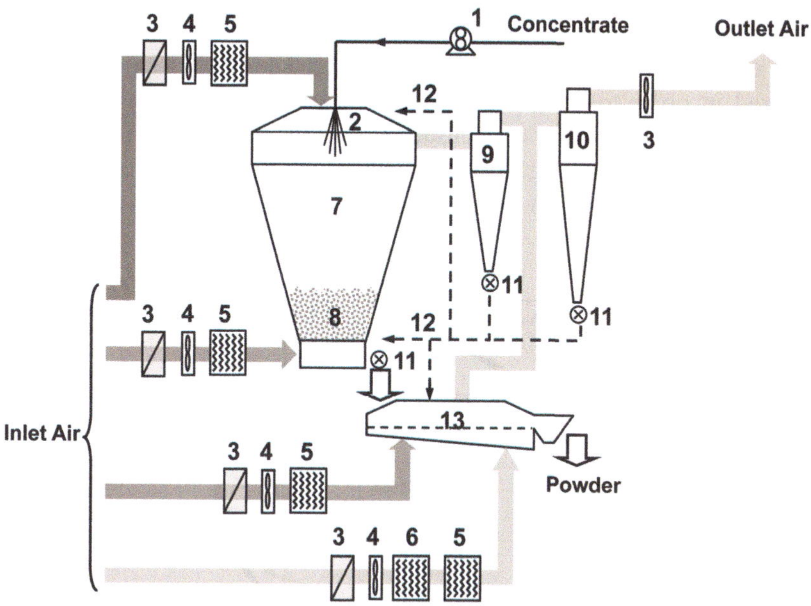

Figure 2.21. *Multistage spray dryer. 1: Feed pump; 2: Sprayer / Air disperser; 3: Air filtration; 4: Air fan; 5: Air heater; 6: Air cooler; 7: Drying chamber; 8: Integrated fluid bed; 9: Primary cyclone; 10: Secondary cyclone; 11: Rotary valve; 12: Reincorporation of fines; 13: Vibro-fluidizer*

There are other dryers such as compact dryers (W dryer), flat-bottom dryers such as "Tall Form", Filtermat®, Paraflash® or the Tixotherm® process. All these dryers have features related to the specific properties of the product (high fat content of fat, starches, maltodextrins, egg products or hygroscopic products).

2.3.3. *Freeze drying*

2.3.3.1. *Principle*

Freeze-drying, or lyophilization, is a dehydration process used to sublimate water in a frozen product: water in the product passes directly from the solid to gas phase (Figure 2.22), bypassing the liquid phase. Sublimation can only occur if the temperature and partial pressure of water vapor are below the triple point. Latent heat of sublimation is almost 2,800 kJ for removing 1 kg of water; the phase transition is endothermic. It is therefore necessary to supply heat in a carefully controlled mannerso as to avoid even partial melting of the frozen water. The water vapor must then be removed in order to maintain the partial pressure of water vapor at a low level. This can be done by condensation at a low temperature (usually as ice) or by using a very dry gas at atmospheric pressure. Figure 2.22 gives a schematic of a freeze dryer equipment.

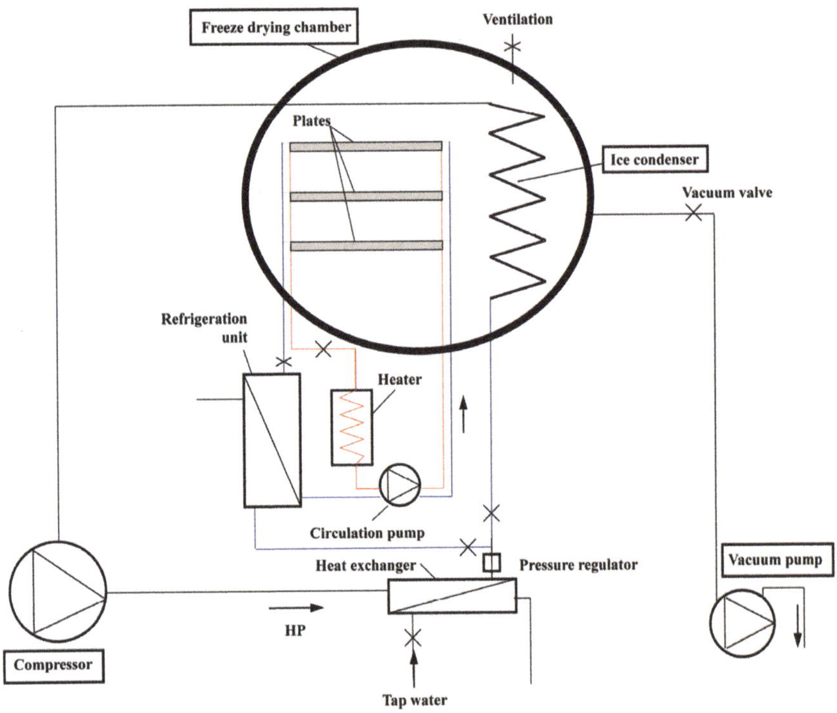

Figure 2.22. *Freeze-dryer*

Vapor transfer always takes place through the dehydrated and porous surface of food. On the contrary, heat transfer to the sublimation front takes place either by conduction through the frozen part or by radiation and/or conduction through the dehydrated layer (Figure 2.23). In the case of rapid heat transfer (freeze-drying at atmospheric pressure with a very dry gas; supply of heat across the frozen layer), it is the transfer of water vapor across the dehydrated layer that limits the overall rate of dehydration. In other cases (vacuum freeze-drying), it is the heat transfer through the dehydrated and porous layer of the food that is the limiting factor.

Freeze-drying is the most suitable process for heat-sensitive products since it preserves the sensory qualities of the food product. However, its cyclic nature and high energy cost are limiting factors to the development of its use.

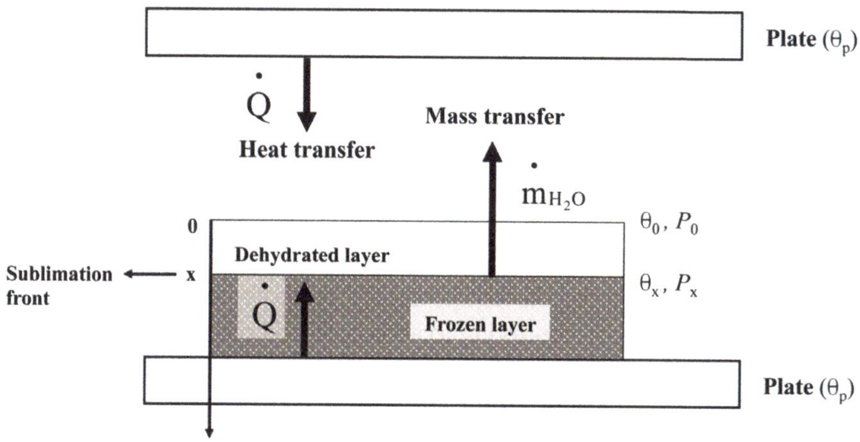

Figure 2.23. *Heat and mass transfers during freeze-drying*

2.3.3.2. *Stages*

Freeze-drying involves five stages (Figure 2.24): freezing (A → B), vacuum (B → C), sublimation (C → D), desorption (D → E) and vacuum release (E → F).

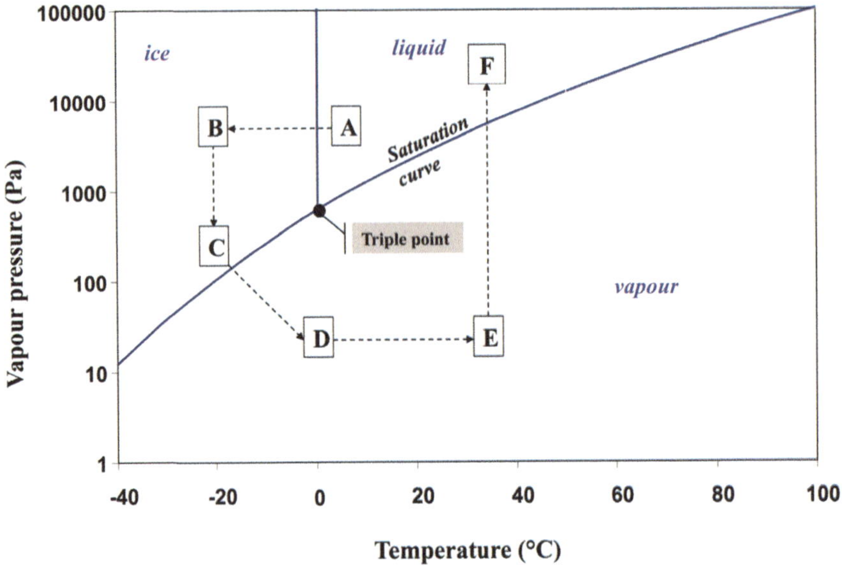

Figure 2.24. *Freeze-drying cycle*

– Freezing (A → B; Figure 2.24): The main advantage of freeze-drying is that dehydration occurs in a solid medium at low temperatures. As a result, there is no movement of liquids or solutes, no volume contraction and virtually no chemical or enzymatic reactions. However, most of the water in the food must be frozen; the temperature must also remain low enough (below –20°C) during freeze-drying to avoid eutectic melting (there is a high risk of this in sugar-rich foods). With some products, it may be useful to avoid freezing the food too rapidly in order to allow larger ice crystals to form, which would increase the pore diameter and facilitate the transfer of water vapor and subsequently rehydration of the freeze dried product for use.

Freezing can either be achieved in a freeze-dryer (internal freezing) by circulating a coolant in the trays, or outside a freeze-dryer (external freezing; see section 2.1):

– Vacuum (B → C; Figure 2.24) is not absolutely necessary. Freeze-drying can be carried out at atmospheric pressure with a stream of cold air or nitrogen provided that the partial pressure of water in the gas is lower than the vapor pressure of ice. Vacuum can however reduce the risk of ice melting and accelerate drying if the limiting factor is the diffusion of vapor through the food product. If the internal heat transfer is the obstacle, as is often the case at the beginning of freeze-drying, it is better to work at low sublimation temperatures and at pressures between 40 and 140 kPa or lower.

Vacuum is often facilitated by a high output pump allowing limited vacuum. This pump is relayed to a second low output pump allowing very low pressures to be achieved. The total duration of the vacuum stage is usually less than 15 minutes.

– Sublimation (C → D; Figure 2.24) corresponds to the primary drying of all frozen water in the food. Its rate is proportional to the difference between the partial pressures of water vapor at the sublimation front and the ice on the condenser. This difference directly depends on the temperature difference between the frozen product and the condenser. With a product at –20°C (P_w ice ≅ 103 Pa) and a condenser at –40°C (P_w ≅ 13 Pa), the vapor pressure difference is 90 Pa; it would be 222 Pa for a condenser at –30°C and a product at –10°C, but with a product at such a temperature, there is a risk of eutectic melting. The relatively small differences in vapor pressure explain the low freeze-drying rates. The dehydration rate decreases as the pressure in the freeze-dryer chamber rises. As a result, vacuum is mostly at an absolute pressure of between 10 and 270 Pa (absolute pressure should ideally be close to a third of the partial pressure of water vapor at the sublimation front). Even under these conditions, the dehydration rate rarely exceeds 1.5 kg m^{-2} h^{-1}. This value corresponds to a shiftin the sublimation front of 2 to 3 cm in 10 hours. In practice, products to be freeze-dried are thinner than this. The thickness of a product plays a key role in the dehydration rate whether it is limited by vapor or heat transfer: the duration of dehydration is approximately proportional to the square of the thickness, such as for the freezing time according to Planck's formula (equation [2.11]) when the limiting factor is the heat exchange within the product.

– Desorption (D → E; Figure 2.24): Desorption corresponds to the secondary drying of bound water. After the primary drying step, there is no more ice in the product and consequently no risk of melting if the temperature rises. The "dry" product temperature increases spontaneously since there is no further sublimation. This temperature increase is necessary for the residual moisture, corresponding to bound water, to be desorbed and evaporate. In practice, the product is maintained at a temperature between 20 and 70°C, under vacuum, for 2 to 6 hours. The product should have a water content corresponding to a monolayer of water, which ensures maximum storage stability (see Volume 1 [JEA 16a], Chapter 1). Controlling the water content during freeze-drying remains challenging on the small scale, but less so in pilot operations.

Figure 2.25 shows that the increase in temperature of the dry layer occurs well before sublimation is finished. Thus, the water content and temperature vary for different product thicknesses. The temperature at the sublimation front remains constant during freeze-drying. During heat transfer through the dry layer, the temperature of the freezing front depends on the temperature of the dry layer, and the risk of the latter overheating is greater than the risk of melting within the frozen part.

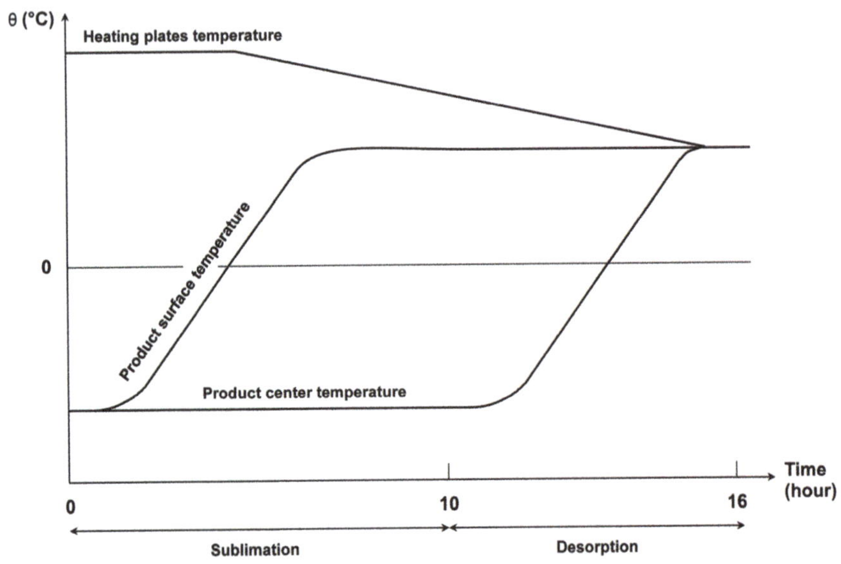

Figure 2.25. *Temperature of the product and heating plates during freeze-drying (heat transfer through the dry layer)*

– Vacuum release (E → F; Figure 2.24): At the end of freeze-drying, the vacuum is generally released by introducing air or nitrogen, which combined with proper packaging protects freeze-dried products from oxidation.

2.4. Stabilization by chemical inhibition

2.4.1. *Preservatives (antibacterial, antifungal)*

The use of preservatives is one of the many methods of ensuring consumer safety, prolonging product shelf life and limiting microbial spoilage. The discoveries of food preservation methods such as fire (for cooking and smoking), salting or bioacidification represent major milestones in human evolution. Salt and smoke constituents were among the first food preservatives used. Today, physical methods (in particular heat) are the most common ways of preserving food due to their effectiveness and ease of implementation. However, the use of chemical preservatives is complementary to these treatments. They can reduce the intensity of physical treatments and limit sensory and nutritional changes [MES 02].

2.4.1.1. *Overview*

A preservative can be defined as any substance incorporated into a food to increase its chemical and microbiological stability and safety. Products traditionally used at food ingredients and having preservation properties (vinegar, sodium chloride, ethyl alcohol, edible oils and sugars) are excluded from this category.

Chemical preservatives should ensure:

– consumer safety by inhibiting the growth of pathogenic microorganisms that may be present (*Salmonella*, *Clostridium*, *Staphylococus*, various fungi) and limiting the production of microbial toxins;

– the sensory stability of food by preventing the growth of spoilage microorganisms.

Preservatives are of benefit when they can substitute costly physical treatments that are difficult to implement. This applies to areas where technical progress is limited due to a lack of training or funding. Additives can be an alternative to the cold chain for example. They also have the advantage of not or hardly modifying the sensory qualities of food and keeping it fresh. As a result, food keeps better and longer, which is desired by the consumer.

Preservatives are mainly antibacterial and their effect is generally slow and limited. They can deteriorate over time due to oxidation or become ineffective by the action of microbial enzymes. In addition, they have a bad image among consumers who question their use. Some preservatives can actually be toxic if consumed in high doses or are suspected of being carcinogenic in the long term. However, most are harmless and occur naturally in fresh or fermented foods. Toxicological studies are necessary however, and regulations are constantly being updated to ensure food safety.

There are two important conditions to ensure effective treatment with preservatives:

– the food product must be of high quality: no preservative can make spoiled food consumable;

– it must be homogenously distributed in the food: preservatives are often used in small quantities and a poor distribution cannot protect the entire product and may cause local overdosing.

Preservatives are often mixed or used in combination with other substances such as antioxidants, which enhances their effectiveness. This facilitates synergism since each family of additives preferentially acts on a group of microorganisms or certain metabolic pathways. Food properties such as pH are very important in ensuring effective treatment. It is therefore crucial to know the exact composition and physicochemical properties of the product, as well as its microflora before formulating a mixture of additives. Moreover, preservatives cannot be used without taking into account any other treatments, primarily physical, carried out on the product.

The use of preservatives in foods is highly regulated, and defined in Europe according to the Directive EC 1333/2008. The principle of positive lists is applied and preservatives can be classified into:

– *generally permitted preservatives*: these include acetic acid, lactic acid, citric acid, malic acid, tartaric acid, ascorbic acid and their sodium salt, potassium and calcium derivatives as well as glucono-δ-lactone and carbon dioxide. These substances, considered additives, are permitted based on professional uses and practices. This general authorization is however limited by a restrictive list of foods in which permitted additives cannot be used. This list concerns raw materials and intermediate products like pasteurised milk or cream, coffee or its extracts, sugars or dried pasta. It is therefore imperative to regularly refer to the regulations, which are complex and constantly changing;

– *conditionally permitted preservatives*: these include sorbic acid, benzoic acid, propionic acid, phosphoric acid and their salt derivatives, fumaric acid, succinic acid, parabens or esters of p-hydroxybenzoic acid, sulphur dioxide (SO_2), sulphites and nitrites.

2.4.1.2. *Salting*

Sodium chloride

Sodium chloride (NaCl), although regarded more as a food ingredient than an additive, is one of the first chemical preservatives to have been used in food. Contrary to once popular opinion, sodium chloride does not act as an antiseptic. It has a weak bactericidal capacity. It is antimicrobial only by its depressant effect on a_w. Salt selects floras depending on their water activity tolerance, inhibits the growth of most spoilage bacteria, but favors the growth of halophiles. It is mainly used to preserve meat, fish and vegetables.

For effective protection, it is necessary to lower a_w values below 0.95 for refrigerated fresh products with a limited shelf life, or below 0.91 for stability outside the cold chain. In the first case, 2 to 3% of salt is necessary; in the second case, 5 to 6% of salt is required after partial drying. Above a concentration of 5%, salt inhibits most anaerobic bacteria and *Pseudomonas* and slows down aerobic bacterial growth.

As regards charcuterie products (cooked or processed meat foods), improvements in the cold chain and hygienic production conditions, the development of vacuum technology and a better knowledge of additives have made it possible to reduce the salt content currently used compared to traditional salting. The salt content of bread is now typically 1.5%, 2% for cooked charcuterie products and 5 to 6% for dried charcuterie products.

Salt is mainly used in cooked, dried and fresh charcuterie products, smoked fish products and pickled vegetables. Salted fish can either be a final product or a raw material intended for further processing: drying, smoking, semi-preserved.

The two existing curing techniques are dry-salting and pickling. In the first case, fish or meat fillets are laid on a bed of fine salt and covered with a thin layer of salt. In the second case, the products are immersed in brine, a concentrated salt solution. Light brine has 16% salt, medium brine has 20% salt and strong brine has 25% salt (close to saturation). Using supersaturated brines helps to maintain the salt saturation of the solution throughout the process depending on its temperature.

With salting, a salt equilibrium between the product and the salt solution is attained in a sufficiently short time to prevent bacterial growth and thereby extend product shelf life. The migration rate of Na^+ and Cl^- ions into the product ("salt penetration") follows Fick's law (equations [1.19–1.20]), and therefore depends on a diffusion coefficient D_m specific to the given matrix, which can be estimated using various experimental methods.

There are two phases in salt penetration:

– the first phase has a high diffusion coefficient, directly related to the water content and a_w of the product;

– the second phase has a low diffusion coefficient, which decreases as the properties of the aqueous phase of the product (viscosity) change.

Salt penetration is therefore rapid at the start and gradually slows down during the salting process. It mainly depends on:

– the quality and size of the salt;

– the quality of the raw material (freshness, type of food, fat content);

– temperature.

Salting negatively impacts food both from a nutritional and sensory perspective; it is important therefore to desalt before consumption.

Nitrites

Nitrites are commonly used salts in the charcuterie industry. However, in Europe its use is prohibited in the processing of fish products. Saltpetre (potassium nitrate) was traditionally used as a preservative. Nowadays, it is often replaced by nitrites, the active form of the salt that inhibits microbial growth. Nitrates act as precursors to nitrites since they are reduced by bacteria in the brine.

The use of nitrates and nitrites is controversial. Nitrites inhibit certain pathogens, in particular *Clostridium botulinum* that may be present in raw foods. Nitrite also has a sensory role in terms of flavor and color: by reacting with myoglobin, nitrosomyoglobin is formed giving meat its stable red color.

Given their toxicity, nitrites are sold mixed with sodium chloride to avoid errors in dosage. They also have the disadvantage of being highly reactive towards amines, forming nitrosamines known to be highly carcinogenic. As a result, nitrite salts are not permitted in the case of fish due to the high levels of endogenous amines.

2.4.1.3. *Sugars*

Consumers primarily associate sugar with sweetness. But sugars have a multitude of other functions that make them important ingredients in many foods. Sugars play an important role in food preservation. The addition of carbohydrates, such as glucose, fructose or sucrose, to jams and jellies prevents microbial growth and preserves food outside the cold chain.

Osmotic dehydration of fruit by dipping in syrup is also a common technique. This results in a reduction in the water content and an increase in the sugar content. Sugar can also be added to tinned vegetables to preserve their texture (firmness) and minimise oxidation responsible for the deterioration of sensory qualities (flavor, color).

Many carbohydrates (glucose, fructose, sorbitol and mannitol) efficiently bind metal ions (copper, iron and, to a lesser extent, cobalt), which can delay catalytic oxidation reactions. In addition, sugars are involved in Maillard reaction and give products antioxidant properties. They can thus be used as additives for some food products (biscuit, sausages).

2.4.1.4. *Smoking*

Smoking, together with salting, is one of the oldest methods of preserving food. Burning certain types of wood can form compounds with bacteriostatic, sensory (color, flavor) and antioxidant properties.

The principle of smoking has largely remained the same. In most cases, the food is pre-salted and slightly dried. The product (meat or fish) is impregnated with smoke from smouldering wood shavings or sawdust. During smoking, the product is dehydrated at the same time as being impregnated with volatile compounds from the smoke as well as undesirable condensed hydrocarbons (benzopyrene).

Nowadays, smoking is mainly used for sensory purposes, at least in industrialised countries. The operation primarily involves giving the product a desired flavor and color, while its preservation is ensured mainly through salting, drying and storing the final product at +4°C (smoked salmon).

The sensory characteristics of smoked fish or meat vary depending on the type of wood used. Hardwoods are most commonly used (beech, oak, walnut, elm, birch). Softwoods should be avoided due to the acidic flavor they impart to products and the substantial contamination by 3-4 benzopyrene that they cause.

Chemically, wood is composed of two parts:

– polyoses: mainly cellulose and hemicellulose (tender part of the wood);

– lignin: hard part of the wood.

In general, the proportion is 50% cellulose and 25% hemicellulose and lignin. The remainder includes a small quantity of complex low molecular weight compounds and a variable amount of water.

Smoke is produced either by burning wood and its components (reactions caused by fire), or by heat generated by the friction of wood causing the main elements to decompose by pyrolysis (chemical decomposition by the action of heat alone). Complete combustion of the wood produces carbon dioxide, water and ash. In order to produce smoke, the reaction must be incomplete.

Pyrolysis of cellulose results in the formation of acetic acid, water and phenols. Pyrolysis of hemicellulose produces aliphatic carboxylic acids. Pyrolysis of lignin produces phenolic compounds.

The composition of smoke is extremely complex and variable depending on the material (type of wood, moisture content, particle size) and pyrolysis conditions (temperature of the fire, air velocity and quantity). It is estimated that smoke has at least 200 constituents; this complexity increases further with oxidation after the introduction of air during smoking. These mostly identified constituents are mainly carbonyl compounds, phenols (most important in the smoking process), organic acids, alcohols, etc. The phenols are mainly guaiacol (2-methoxyphenol), eugenol (4-allylguaiacol) and vanillin. The amount of phenols and acids increases with the amount of oxygen supplied. The level of guaiacol is often used as a standard to estimate the intensity of smoking.

Smoke contains polycyclic aromatic hydrocarbons (tars), some of which are carcinogenic (mainly 3-4 benzopyrene). These products are generally more associated with the smoke particles than the gas phase. The level of 3-4 benzopyrene can be limited with a pyrolysis temperature of 450°C; with cold smoking, absorption in fish is eight times higher compared to hot smoking. The World Health Organization (WHO) recommends a maximum content of 1 μg kg^{-1} of 3-4 benzopyrene, which is rarely exceeded in EU.

Burning or pyrolysis of wood produces a range of gases and vapors that condense into a stable aerosol in the coldest part of the facility. This aerosol is composed of a gas phase and a particle phase (fine droplets), which form the visible part of smoke; the compounds in these phases are the same only in different concentrations. The absorption of different smoke constituents and their penetration into the product depends on several factors. The results vary based on the techniques used and the aim of the technological treatment.

Smoking is a physical process based on three phenomena:

– adsorption (and concurrent desorption of water vapor);

– dissolution;

– diffusion.

The adsorption of smoke constituents is considerably higher on a wet surface than a dry surface, whereas absorption of the particle phase is conversely higher on a dry surface. Therefore, the best conditions for adsorption are at the beginning of the smoking process when absorption of the particle phase is negligible. The rate of penetration of phenolic compounds is a complex function of the concentration and composition of smoke, environmental conditions (temperature and relative humidity) and the nature of the product surface.

The smoke constituents adsorbed on the surface then diffuse into the interior of the flesh, which is referred to as "smoke penetration". In the case of smoking, the concentration on the surface layer of the product constantly varies with time, and certain lipophilic components are distributed in the lipid phase. For these reasons, the equation on the diffusion of matter (Fick's law) cannot be applied.

Penetration of smoke constituents continues during storage depending on the temperature.

The main parameters that modify the dynamics of smoking are smoke density, temperature, the moisture content of the product, and the relative humidity and velocity of air.

Sensory role (flavor, color, texture)

The factors responsible for flavor are still poorly understood. Typical aromas appear to be due to phenols (smokey smell), but carbonyl compounds and acids are responsible for the various flavors. Syringol is typically responsible for the smokey smell while guaiacol and eugenol are responsible for the smokey taste. The type of wood therefore significantly impacts flavor.

The color of smoked fish or meat is mainly due to Maillard reaction products. The color varies depending on the wood used, which can be attributed to the color of the different carbonyl and phenolic compounds.

Changes in texture are mainly due to the combined action of salt (which increases the vapor pressure of volatile components), heat and certain smoke constituents, which can cause protein coagulation (formation of a bright shiny film on the surface: gelatinization of the subcutaneous layer). As a result, there are differences between cold and hot smoked products. Cold smoking produces a soft and tender flesh texture whereas hot smoking produces a harder and drier texture.

Antioxidant role

Smoke has an antioxidant effect due to the capacity of phenols (syringol, guaiacol, eugenol, vanillin, etc.) to form stable free radicals, which limits the development of a rancid flavor.

Antimicrobial role

In hot smoking, heat plays the main role in the destruction of microorganisms whereas with cold smoking, smoke is the active component. Smoke can have an antiseptic role due to phenolic compounds at low boiling point, but this action is limited. The microbiological stability of smoked products depends on the combined action of smoke, heat and salt, which lowers the a_w.

2.4.1.5. *Other antiseptic molecules*

Weak acids

Weak acids are an important group of preservatives that can inhibit the growth of microorganisms and limit the germination of bacterial spores. This bacteriostatic effect is linked to specific properties and/or weak acid functions. There are three types of effects:

– *decrease in pH*: acids acidify the medium (e.g. lactic acid);

– *reduced ionization of weak acids due to a drop in pH*: the undissociated form of these acids can act against microorganisms. Unlike H^+ ions, undissociated molecules passively diffuse into the intracellular environment where they modify the pH by dissociating and releasing H^+ ions. The proportion of undissociated acid increases as the pH drops; therefore weak acids are most active at acidic pH, below pH 5 to 6 depending on the pKa of the acids. Changing the homeostatic balance of the cytoplasm by the penetration of weak acids can alter and even rupture the plasma membrane, which leads to impaired cell viability. In addition, enzymatic reactions have an optimum pH above or below which their kinetics is affected. Any variation in the pH of the cytoplasm will therefore lead to slower enzymatic activity and growth, and disrupt metabolic pathways. However, not all enzymes exhibit the same pH dependency; extracellular hydrolases in particular are less sensitive than cytoplasmic enzymes;

– *specific toxic effects*: certain acids such as lactic acid, acetic acid or citric acid are used in concentrations of about 1% or more. Their toxicity is variable and linked mainly to the acid function. Other acids (benzoic acid, propionic acid and sorbic acid), sulphites or nitrites are effective at lower concentrations of between 0.05 and 0.2%. They act on specific metabolic pathways, e.g. sulphites act on proteins, and sorbic acid limits the availability of α-unsaturated fatty acids.

Other preservatives

Stabilizing food by inhibiting microbial growth using preservatives is usually part of a process involving other factors. Preservatives can be mixed or used in combination with physical processes (synergistic effect).

Tables 2.5, 2.6 and 2.7 summarize the properties of the main preservatives used in the food industry. The maximum allowable concentration of each preservative is specified for each food.

Category	*Examples*	*Properties*	*Typical use in food products*
Phosphates	Phosphoric acid Sodium, potassium and calcium salts	Antimicrobial	Non-alcoholic beverages Sterilised and UHT milk Cream, processed cheeses Meat products Soups and broths Liquid eggs Cider Powders Surimi Potato products
Nitrates and nitrites	$NaNO_3$; KNO_3 $NaNO_2$; KNO_2	Anti-*Clostridium* Synergistic effect with drop in pH and a_w Pink color of processed meat products	Raw or cooked cured meat products Canned meat Foie gras
Sulphur dioxide and sulphites	Sulphur dioxide Sodium sulphite Sodium bisulphite Sodium or potassium metabisulphite Calcium sulphite Calcium hydrogen sulphite Potassium bisulphite	Antimicrobial (bacteria, fungi) Inhibits enzymatic browning or color instability Whitens cod flesh	Alcoholic or non-alcoholic beverages made from fruit, grain or honey Vinegar and mustard Glucose syrups and other sugar products
Carbon dioxide and carbonates	CO_2	Fungistatic and bacteriostatic (more effective against Gram-negative) Increases product shelf life	Fizzy drinks Meat, cakes and pastries, ready meals (modified atmosphere)

Table 2.5. *Mineral preservatives (except chlorides)*

Molecules	*Properties*	*Typical use in food products*
Sorbic acids, sorbates (potassium and calcium)	Fungistatic Less antibacterial Synergistic inhibitory effect with other preservatives Active up to pH 6.5 (useful with weak acids)	Non-alcoholic beverages Jams Fruit and vegetables Semi-preserved and dried fish Fish eggs, shrimp Processed cheese Liquid and dried eggs Confectionary, sauces, mustards and seasonings Non-sterilised soups
Benzoic acid and benzoates (sodium, potassium and calcium)	Effective against bacteria, yeasts and to a lesser extent fungi Active up to pH 4.5 (weak acids)	Alcoholic or non-alcoholic beverages Vegetables in brine Candied fruit Semi-preserved fish and fish eggs Chewing gum Sauces, prepared salads, mustard
Parahydroxybenzoic acid esters (parabens)	Effective against fungi, yeasts, and to a lesser extent bacteria Synergy between parabens and essential oils Active up to pH 8	Jelly coating for meat products, surface treatment for dried meat, grain and potato snacks, confectionary
Propionic acid and propionates (sodium, calcium and potassium)	Active against fungi pH < 5	Bakery products (sliced and pre-packed bread) Industrial pastries

Table 2.6. *Organic preservatives*

Molecules	*Properties*	*Typical use in food products*
Dimethyl dicarbonate (DCDM)	Very effective product Inhibits yeasts Also effective against fungi and bacteria	Wine Ready-to-drink teas Sports drinks Fruit juices
Biphenyl, orthophenylphenol, thiabendazole	Antifungal	Surface treatment of citrus fruit and bananas
Nisin	Bacteriocin produced mainly by *Lactococcus lactis* Effective against *Clostridium butyricum* Less active against *Escheridia coli* and *Listeria monocytogenes* (unless synergistic effect with a sublethal heat treatment)	Protects cheese from *Clostridium butyricum* Semolina and tapioca Ripened and processed cheeses Clotted cream
Natamycin	Antibiotic Effective against several types of fungi	Surface treatment of hard, semi-hard and soft cheeses Dry sausage and salami
Calcium disodium ethylenediamine-tetraacetate (EDTA)	Antibacterial by its weak acid function Chelating agent for divalent ions Decreases the heat resistance of *C. sporogenes* at pH 7	Emulsified sauces Preserved crustaceans and molluscs Preserved fish Frozen and deep-frozen crustaceans Preserved vegetables, mushrooms and artichokes

Table 2.7. *Other preservatives*

2.4.2. *Fermentation*

Fermentation is a biological stabilization method that has been used for more than 6,000 years. The most common types of fermentation are alcoholic, lactic and acetic acid fermentation, which form the basis of wine, beer, cheese and vinegar. Fermentation increases the nutrient content of food through the biosynthesis of vitamins, essential amino acids and proteins, improves protein and fibre digestibility, enhances mircronutrient bioavailability and reduces anti-nutrient factors. Fermented foods also have therapeutic properties (pro- and prebiotics).

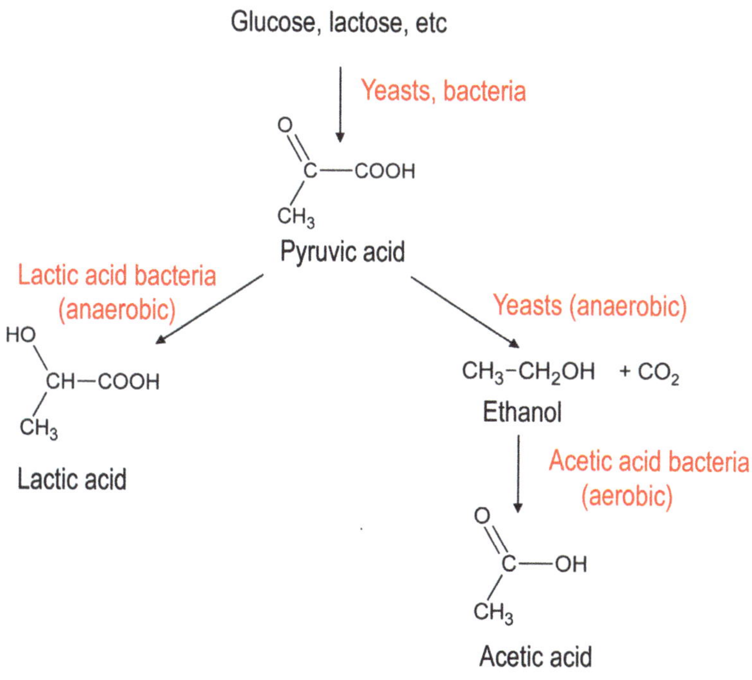

Figure 2.26. *Main fermentative pathways*

Fermentation improves food safety by reducing the risk of toxic compounds developing such as aflatoxins and cyanogens, and by producing antimicrobial agents like lactic acid, bacteriocins, carbon dioxide, hydrogen peroxide and ethyl alcohol, which facilitate the inhibition or elimination of pathogens. These fermentation products are inhibitors above a certain concentration, whether it be alcohol, lactic acid or acetic acid. Fermentation can cause physicochemical changes such as a drop in pH, a reduction in the redox potential, etc., which may inhibit the development of certain flora; carbon or nitrogen substrates and growth factors (amino acids, vitamins) may be depleted in certain products, which deters the growth of undesirable flora.

Apart from its nutritional value and stabilising role, fermentation enriches food by producing a variety of flavors, textures and aromas.

Two types of microorganisms are involved in fermentation:

– bacteria, which play a major role in lactic and acetic fermentation;

– yeasts, which are mainly involved in alcoholic fermentation, and moulds, useful in the manufacture of certain cheeses (Roquefort, Camembert).

Fermentation takes place when the temperature, acidity, oxygen and nutrient conditions are favorable to microbial growth. They then synthesize enzymes needed to break down macromolecules into smaller molecules, which can then be assimilated. A complex series of chemical reactions occurs resulting in the final product:

– alcoholic fermentation converts sugars into alcohol and carbon dioxide mainly through the action of yeast. This type of fermentation is anaerobic (in the absence of oxygen). It is used in the manufacture of beverages (wine, beer) and bread;

– lactic acid fermentation, also anaerobic, converts sugars into lactic acid. This is done mainly through the action of lactic acid bacteria. This type of fermentation is used in the manufacture of several foods (dairy products, cooked and processed meat products, preserved vegetables, etc.);

– acetic acid fermentation occurs in the presence of oxygen (aerobic fermentation). Acetic bacteria convert alcohol into acetic acid, the main component in vinegar;

– other types of fermentation (propionic, butyric, malolactic, etc.).

During a fermentation process, different species of microorganisms may be involved. Two different types of fermentation may occur simultaneously: lactic acid and alcoholic fermentation in the manufacture of Kefir from milk; or successively: lactic acid fermentation followed by propionic acid fermentation in cheese-making, alcoholic fermentation followed by malolactic acid fermentation in wine-making (see Volume 3 [JEA 16b]).

3

Separation of Food Modifying Agents

This chapter focuses on processes based on the non-destructive separation of microorganisms in liquids by differences in the density or size. Centrifugal sedimentation and cross-flow microfiltration are the most common separation methods.

3.1. Sedimentation

Separation by sedimentation may be carried out when the dispersed elements (microorganisms, fat globules, insoluble proteins, lactose crystals, glycerides) have a different density to the liquid or dispersing phase. It is mainly used in the dairy sector to separate butyric spores from milk used in the manufacture of hard pressed cheeses (bactofugation).

3.1.1. *Stokes' law*

The study of the movement of a spherical particle in a fluid under the action of a force was described in Chapter 1, section 1.2.5.2. In practice, the separation of such particles by sedimentation always takes places in laminar flow given the small size (diameter *D*), that is for Reynolds number (Re) values between 10^{-3} and 2. In this case, the relationship linking the Newton and Re numbers is:

$$\mathrm{Ne} = 24\,\mathrm{Re}^{-1} \qquad [3.1]$$

Chapter written by Romain JEANTET.

Hence:

$$\frac{F_M}{\pi \frac{D^2}{4} \frac{v^2}{2} \rho_f} = 24 \frac{\eta}{D \rho_f v} \qquad [3.2]$$

D is the particle diameter (m), η and ρ_f are the viscosity (Pa s) and density (kg m^{-3}) of the fluid, respectively, and v is the particle's velocity (m s^{-1}).

In sedimentation, it is assumed that the duration of the transient phase, to reach maximum velocity, is negligible. It is therefore accepted that sedimentation occurs in steady state, that is at a constant speed, and the driving force F_M is therefore equal to the resistance to the movement F due to the friction in the fluid. Where a spherical particle moves due to the action of gravity, F_M is the resulting force of weight and buoyancy (Figure 3.1), namely:

$$F_M = \frac{\pi D^3}{6} (\rho_p - \rho_f) g \qquad [3.3]$$

where ρ_p is the particle density (kg m^{-3}) and g is the gravitational acceleration (9.81 m s^{-2}).

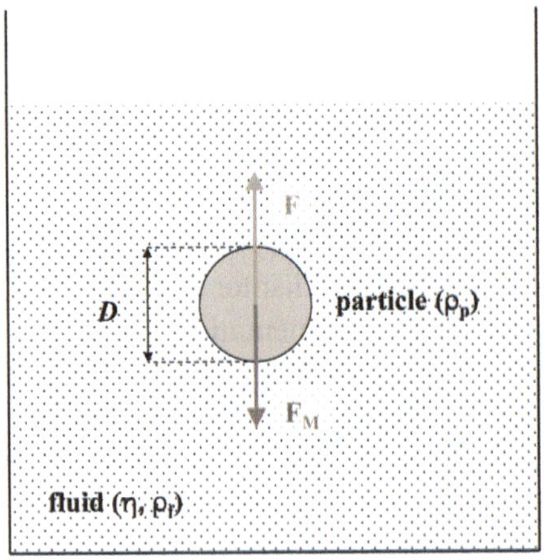

Figure 3.1. *Sedimentation of a spherical particle under the influence of gravity*

By combining [3.2] and [3.3], we obtain the Stokes' relation:

$$\mathrm{v} = \frac{D^2}{18\eta}(\rho_\mathrm{p} - \rho_\mathrm{f}).\mathrm{g} \qquad [3.4]$$

Generally, the density difference between the particle and the dispersing phase ($\Delta\rho = \rho_p - \rho_f$) varies little with temperature. However, temperature strongly affects the viscosity of the dispersing phase. This effect is utilized during the skimming of milk at temperatures between 50 and 60°C where viscosity is between 0.9 and 1×10^{-3} Pa s as opposed to 2×10^{-3} Pa s at 20°C.

Using equation [3.4], it is possible to determine the velocity of a fat globule (creaming; $\Delta\rho = -117$ kg m^{-3}; $D = 2\times10^{-6}$ m) and a *Clostridium butyricum* spore (sedimentation; $\Delta\rho = 240$ kg m^{-3}; $D = 0.5\times10^{-6}$ m) in milk at 20°C. These are respectively equal to 1.3×10^{-7} m s^{-1} and 1.6×10^{-8} m s^{-1}, corresponding to a ratio of 1 to 8. The higher $\Delta\rho$ for spores does not compensate for the smaller size (D), as the latter contributes as D^2 in Stokes' law.

3.1.2. *Centrifugal sedimentation*

According to Stokes' equation, it appears that the only factor that can increase the sedimentation velocity is a temperature increase, which lowers the viscosity of the dispersing liquid phase. However, temperature increases are limited due to the thermal sensitivity of the product constituents.

In order to increase the sedimentation velocity, the only remaining factor that can be changed is the acceleration. This is used in sedimentation techniques under the influence of a centrifugal acceleration field. Centrifugal separators are used to separate light phases (fat) from heavy phases (impurities, aggregated proteins, microbial spores, etc.). Bactofugation requires rotational speeds 1.5 to 2 times greater than those for skimming. It reduces flora by 1 to 1.5 log at nominal flow. This efficiency can be improved by lowering the flow rate (increase in residence time and consequently sedimentation) or carrying out double bactofugation.

Figure 3.2 shows a diagram of a centrifugal separator without plates. The position of the interface between the two phases can be adjusted by varying

R_1 and R_2, which controls the purity of the separated phases. For example, the level of separation approaches the axis if R_1 is reduced. This property is useful in completely separating the heavy phase (e.g. oil separator in an oil treatment plant); however, due to the proximity of the interface with the discharge of the light phase, there is a considerable risk of pollution by the heavy phase. A regenerator is used in this case. On the contrary, if the aim is to obtain a purified light phase, the interface position is moved away from the axis: a purifier is used in this case.

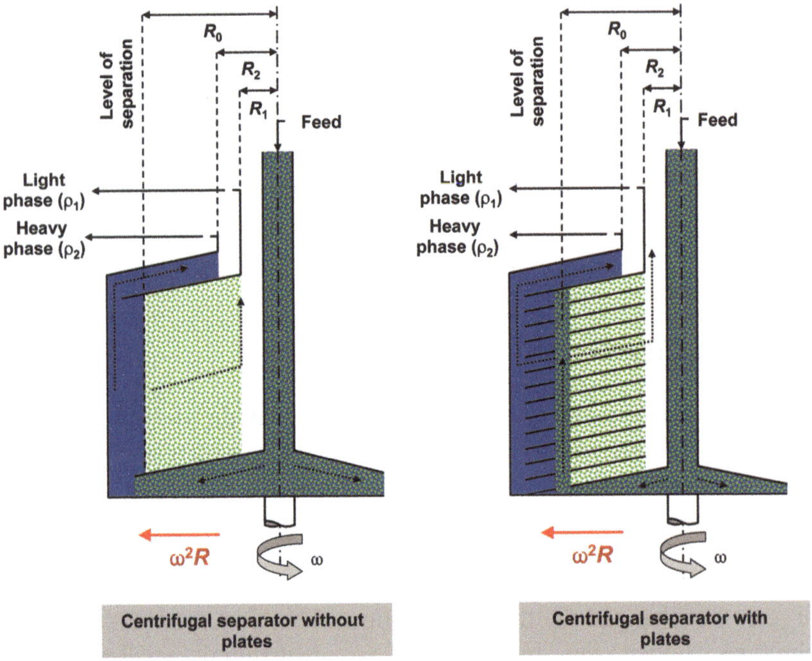

Figure 3.2. *Diagram of a centrifugal separator with or without plates*

In order to increase the efficiency of a centrifugal separator, truncated conical plates are attached inside the bowl (Figure 3.2). This system has many advantages:

– adding plates decreases the travel distance of the dispersed elements, and thus facilitates their separation since they are considered separated once they reach the plate walls (effect of agglomeration with other particles; according to Stokes' law, the sedimentation velocity increases with the square of the diameter of the agglomerated particles);

– a liquid with a mass m at position R_0 should receive kinetic energy of $\frac{1}{2}$ m ω^2 R_0^2, supplied by the centrifuge engine. This mass of liquid is separated in two parts with the same angular velocity ω, and positioned in R_1 and R_2 both lower than R_0. These parts, therefore, have less kinetic energy than the incoming liquid. In a separator without plates, there is a tendency for the liquid to accelerate, causing swirls which disrupts the sedimentation process. Conversely, by increasing the contact surface between the centrifuge bowl and the liquid, the presence of plates causes an entrainment of the bowl and consequently the recovery of the kinetic energy;

– plates have the effect of reducing the hydraulic diameter D_h of the passage, which results in Re values consistent with laminar flow.

A particle located between two plates tends to move with the dispersing phase at velocity v_p, and to be deposited centrifugally at velocity v_c, as expressed by Stokes' law:

$$v_c = \frac{D^2}{18\eta}(\rho_p - \rho_f)\,\omega^2 R \qquad [3.5]$$

where R is the distance from the axis of rotation (in m) and ω the angular velocity (in rad s^{-1}). This velocity increases the further the particle moves from the axis of rotation. Conversely, the value of v_p decreases as R increases. Where v_p decreases and v_c increases as R increases, the resultant velocity vector v_r gradually moves towards the plate (Figure 3.3).

3.2. Cross-flow filtration

Cross-flow filtration is carried out at various levels in the treatment of food liquids and effluents. The different fields of application of these techniques (microfiltration (MF), ultrafiltration (UF), nanofiltration (NF) and reverse osmosis (RO)) are outlined in Figure 3.4.

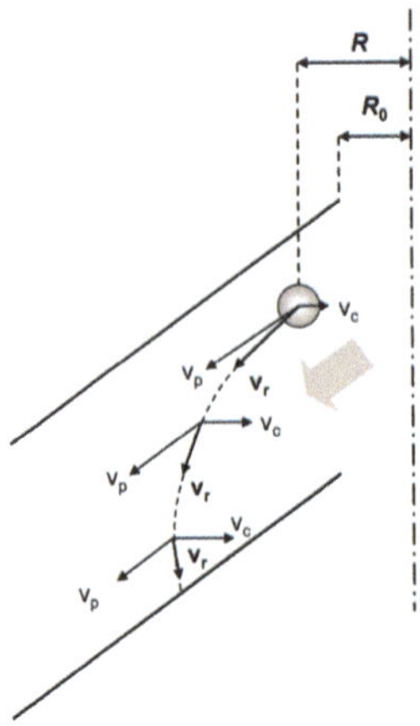

Figure 3.3. *Change in resultant velocity v_r as a function of position*

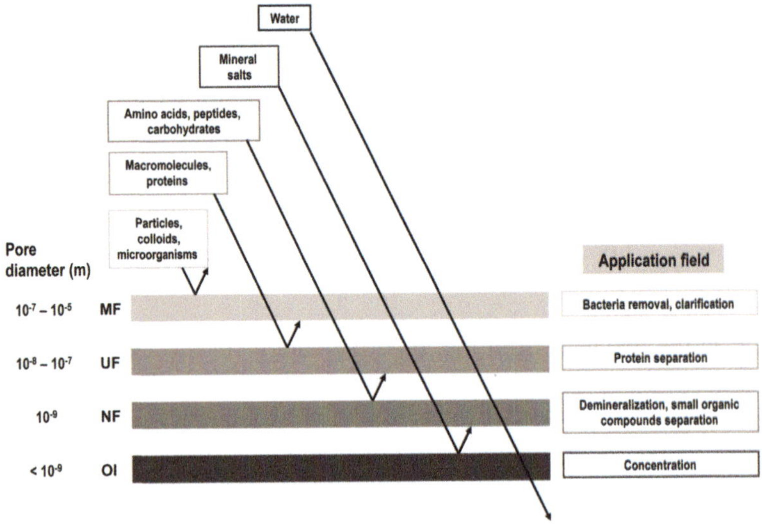

Figure 3.4. *Cross-flow filtration processes*

Of these processes, microfiltration can selectively concentrate the dispersed elements, including microorganisms. Thus, it can be used for undesired bacteria (spoilage and/or pathogenic) removal. In this case, the transfer of solvent and solutes through the membrane occurs by entrainment in the pores.

3.2.1. *Solvent transfer laws*

Darcy's law

If the pores of a membrane are compared to a set of capillaries arranged parallel to each other with a length l, corresponding to the thickness of the membrane, and diameter $D = 2R$, the volume flow rate of a pore according to Poiseuille's law (equation [1.28]) is:

$$\dot{V} = \frac{\Delta P}{8\eta\, l} \pi R^4 \qquad [3.6]$$

where ΔP is the pressure drop in the pore (also known as the transmembrane pressure *TMP*, in Pa), and η is the viscosity of the filtrate that flows through the membrane pores. Given an average of n pores per unit area, the flow rate per unit area, also known as the permeate flux density J (in m s^{-1}, often expressed as l h^{-1} m^{-2}), is equal to:

$$J = \frac{n\ \pi\ R^4}{8\,l} \frac{TMP}{\eta} \qquad [3.7]$$

This expression can be written as the more familiar Darcy's law:

$$J = \frac{1}{R_m} \frac{TMP}{\eta} \qquad [3.8]$$

R_m is the hydraulic resistance of the membrane (m^{-1}), which depends thus (equation [3.7]) on the pore diameter, pore density and membrane thickness. This law applies when filtering a Newtonian fluid (e.g. water or air), which is free of any particles that could be retained on the membrane. It is interesting to note that J is a function of three terms: the first corresponds to the operating conditions (ΔP), the second to the membrane properties (R_m) and the third to the fluid flowing through the pores (which viscosity is η).

During the filtration of solutions containing macromolecules or dispersed elements, there is an accumulation of highly concentrated material on one side of the membrane, which causes additional hydraulic resistance (R_p) also known as the concentration polarisation phenomenon. During the process, the flux density can decrease due to fouling (deposits), which also creates further resistance (R_f).

Thus, permeation flux density that takes into account these different hydraulic resistances is known as "hydraulic resistance in series":

$$J = \frac{1}{(R_m + R_p + R_f)} \frac{TMP}{\eta} \qquad [3.9]$$

With solvent transfer, just like with electrical current or heat transfer $(\frac{1}{h})$, resistance in series adds up. The ratio $\frac{1}{\eta(R_m + R_p + R_f)}$, which is inverse to the overall hydraulic resistance to solvent transport, represents membrane permeability (m s^{-1} Pa^{-1}).

Film model

The concentration C_m of a solute S at the membrane depends on a mass balance on the boundary layer at the membrane taking into account two mass fluxes: the solute transport by convection towards the membrane (J $C_{(S)}$) and the diffusion in the opposite direction due to the concentration gradient $(-D_m \frac{\partial C_{(S)}}{\partial x})$ (Figure 3.5).

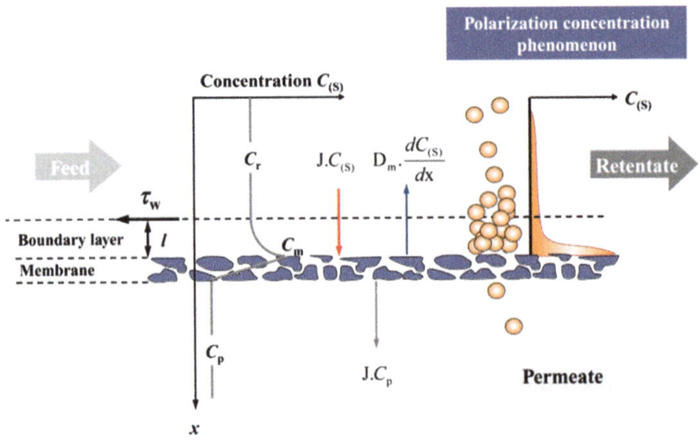

Figure 3.5. *Film model*

After reaching steady-state, the following mass balance applies per unit area:

$$J\,C_p = J\,C_{(S)} - D_m \frac{\partial C_{(S)}}{\partial x} \qquad [3.10]$$

where D_m is the diffusion coefficient of the solute ($m^2\ s^{-1}$), and C_p is the solute permeate concentration ($kg\ m^{-3}$). By integrating equation [3.10] over the boundary layer thickness l (m), we obtain:

$$J = \frac{D_m}{l} \ln\left(\frac{C_m - C_p}{C_r - C_p}\right) \qquad [3.11]$$

where C_r et C_m are the respective concentrations of the solute A in the retentate and at the membrane surface ($kg\ m^{-3}$). Sometimes $\frac{D_m}{l}$ ($m\ s^{-1}$) can be replaced by k, which is the mass transfer coefficient of the solute (equation [1.20]). This phenomenon called concentration polarization is reversible, which is why the concentration polarization (i.e., boundary) layer is sometimes referred to as the dynamic layer. However, this phenomenon may lead to irreversible deposit when the concentration C_m exceeds a critical concentration (solubility limit of mineral salts, sol gel transition of proteins, etc.).

More generally, solute transport by convection to the membrane and back to the liquid stream by diffusion/erosion can be characterized in cross-flow filtration by a dimensionless number, $\frac{\rho\ J^2}{\tau_w}$, where ρ is the density of the filtrate and τ_w is the wall shear stress. Despite this, research in this area has more frequently focused on the ratio $\frac{J}{\tau_w}$, which directly reflects the competition between convective transport (proportional to J) and the diffusion/erosion of the deposit (proportional to τ_w). Qualitatively, this ratio addresses the same phenomenon, but has the disadvantage of not being dimensionless (unit: $m\ s^{-1}\ Pa^{-1}$).

3.2.2. *Influence of filtration parameters*

Transmembrane pressure *TMP* and wall shear stress τ_w are the main parameters influencing permeation flux density J.

As shown in Figure 3.6, J levels off as *TMP* increases. This value increases with rising wall shear stress τ_w, which, all other parameters being equal, is linked to an increase in the flow rate. Pressure causes compaction of the membrane, the concentration polarization layer and any deposits formed. Compaction of these layers, according to equation [3.9], results in an increase in corresponding hydraulic resistance. Beyond a certain limit, the influence of compaction on the increase in overall hydraulic resistance to solvent transfer becomes dominant on the increase in pressure: J reaches a maximum and levels off.

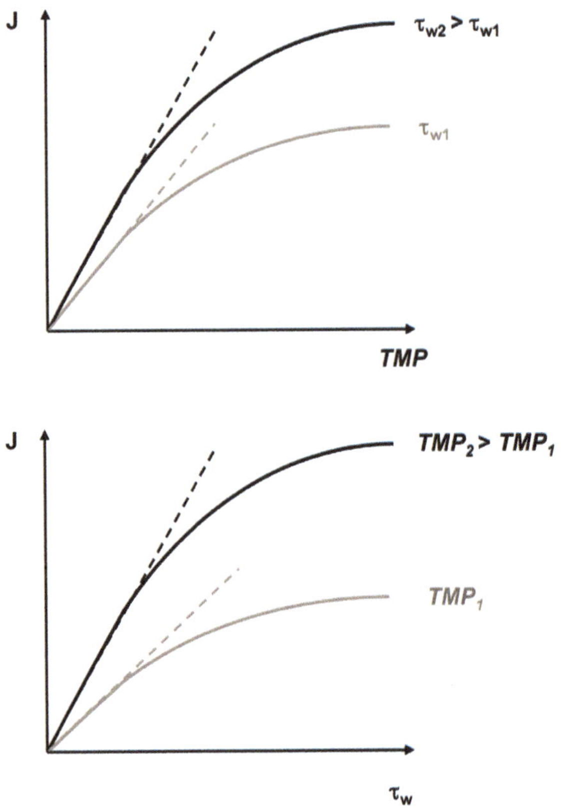

Figure 3.6. *Influence of transmembrane pressure and wall shear stress on permeation flux density*

In contrast, an increase in τ_w results in a decrease in thickness of the boundary layer of the membrane and increased erosion of deposits formed, which lowers the overall hydraulic resistance to solvent transfer: the maximum value of J is higher in this case.

The use of filter modules can be optimized as follows: for a fixed *TMP* value, there is an optimal τ_w value in terms of the productivity, operating time and total cost (depreciation and operation) of the process.

Transmembrane pressure varies along the membrane wall if the permeate-side pressure P_P is the same across the entire surface (static permeation):

– at the inlet, $TMP_I = P_I - P_P$

– at the outlet, $TMP_O = P_O - P_P$

where P_I and P_O are, respectively, the pressure of the retentate at the inlet and outlet of the membrane module. The difference $[P_I - P_O]$ corresponds to the pressure drop ΔP, which depends on the membrane type, flow rate, viscosity and density of the product (see Chapter 1, section 1.2.5.3). Average transmembrane pressure is therefore:

$$TMP_A = \frac{P_I + P_O}{2} - P_P \qquad [3.12]$$

Cross-flow filtration is usually carried out at pressures greater than the pressure drop, or $\Delta P = TMP_I - TMP_O = P_I - P_O \ll TMP_A$. However, in microfiltration, the dynamic layer can play a crucial role in the performance and selectivity of the membrane. The formation of this layer must be controlled, which explains the need to work at low *TMP* (< 0.1 MPa). In this case, given the module pressure drop, the transmembrane pressure can be greater than 0.1 MPa at the inlet, and very low (even negative) at the outlet. To overcome this difficulty, the "Bactocatch®" concept was developed (Figure 3.7): it involves creating an identical pressure drop on both sides of the membrane by ensuring a co-current flow of permeate (dynamic counter pressure). A comparison of both systems is shown in Figure 3.8.

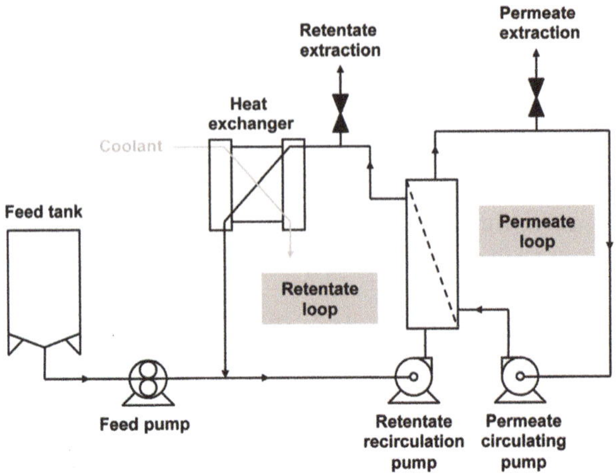

Figure 3.7. *Continuous filtration unit: Bactocatch system*

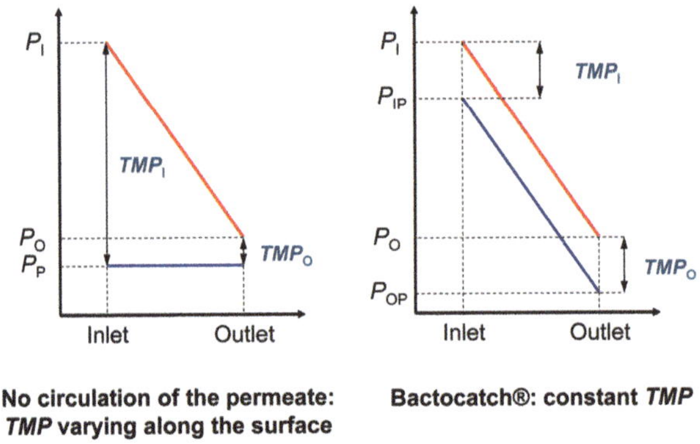

Figure 3.8. *TMP value along the membrane surface with or without dynamic counter-pressure (Bactocatch). P_I, P_O: retentate pressure at the inlet and outlet of the membrane module, respectively; P_P : permeate-side pressure; P_{IP}, P_{OP}: permeate pressure at the inlet and outlet of the membrane module, respectively*

Filtration can be characterized by the volume reduction ratio (VRR) used. The VRR is expressed as follows:

– in batch mode.

$$VRR = \frac{V_f}{V_r} = \frac{m_f \, \rho_r}{m_r \, \rho_f} \qquad [3.13]$$

m_f, V_f, ρ_f and m_r, V_r, ρ_r are, respectively, the masses, volumes and densities of the feed and the retentate;

– in continuous mode

$$VRR = \frac{\dot{V}_f}{\dot{V}_r} = 1 + \frac{\dot{V}_p}{\dot{V}_r} \quad [3.14]$$

$\dot{V}_f$, $\dot{V}_r$ and $\dot{V}_p$ are the volume flow rates of treated liquid, retentate and permeate.

The permeation flux density decreases as the VRR increases, which is due to an increase in the concentration of the solute or dispersed elements. This drop in performance requires the use of multi-stage facilities.

In addition, there is a reduction in filtration performance over time due to fouling, which results in an increase in R_f (equation [3.9]). In order to maintain performance, *TMP* and τ_w could in this case be increased within certain limits since the energy cost and R_p resistance increase with *TMP* and τ_w.

An increase in the temperature changes the characteristics of the dynamic layer and reduces the viscosity of the filtrate. However, above 60°C, the effect on viscosity is limited and a risk of protein denaturation and mineral clogging exists (calcium phosphate precipitation during the filtration of dairy liquids).

Microfiltration is currently used in the dairy industry to remove microorganisms. It requires prior skimming of fluids (milk, whey) and can reduce the cell count by 3 to 4 log. Applications are being developed in the brewing (to replace Kieselguhr filtration) and wine industries.

4

Inactivation of Food Modifying Agents

4.1. Heat treatment

The stabilization of food using heat is widely used in the food sector and serves several purposes:

– it partially or totally destroys spoilage flora (*Micrococcus*, *Bacillus,* psychrotrophic flora, lactic flora, etc.) and pathogenic or toxigenic flora (*Salmonella,* *Staphylococcus, Clostridium perfringens* or *botulinum*) to improve the safety and quality of food products;

– it inactivates certain enzymes, either endogenous (plasmin, lipoxygenase, polyphenoloxidase) or from microorganisms (microbial lipases) to improve product stability during storage.

Increasing the temperature of a medium causes the rupture of low energy bonds (hydrogen or ionic bonds), which changes the conformation of some macromolecules (DNA, intracellular enzymes, wall proteins, etc.). This results in a drop in biological activity leading to enzyme inactivation and cell death. A heat treatment based on these two objectives can extend a product's shelf life by several days to several years. Moreover, heat treatment can:

– change the structure of certain compounds such as proteins, as well as their physicochemical and functional properties: for example, certain proteins can be denatured to improve texture and yield (yoghurt, cheese or surimi);

Chapter written by Romain JEANTET.

– generate favorable Maillard reaction products in some applications (chocolate and bread) or avoid Maillard reaction in others (UHT milk).

In general, the reasoning behind heat treatment is threefold: reduce biological activity (with the aim of maximizing this), influence the sensory and nutritional quality of food products (based on product specifications) and treatment cost (with the aim of minimizing this).

There are two basic treatment categories:

– *pasteurization* involves the destruction of only the vegetative forms of microorganisms at temperatures between 60 and 100°C, including common flora and certain pathogens (*Salmonella*, *Brucella*, *Listeria*, etc.). It can extend the shelf life of chilled products by several days to several weeks: in this case, reference is made to the "use-by date".

– *sterilization* corresponds to the destruction of all microorganisms in the product at temperatures above 100°C, including spore forms. The resulting products have a "best before date" ranging from several months to several years at room temperature.

In the following, the kinetics of destruction of microorganisms and modification of components will be dealt with first followed by the different implementation methods.

4.1.1. *Destruction of microorganisms at constant temperature*

4.1.1.1. *Kinetics of microbial destruction at constant temperature*

When an aqueous suspension of a pure strain of microorganisms at the same stage of growth (in particular, entirely vegetative or spore forms) is maintained at a lethal temperature θ (i.e. causing microbial mortality), cell death follows first-order kinetics:

$$\frac{dN}{dt} = -k_\theta(N) \qquad [4.1]$$

N is the number of microorganisms at time t and k_θ is the rate constant (s^{-1}), which depends on the microorganism, temperature θ and physicochemical characteristics of the medium (pH, composition, a_w, etc.). This equation is integrated between population N_0 at time 0 and population N at time t as follows:

$$\log\left(\frac{N_0}{N}\right) = \frac{t}{D_\theta} \quad [4.2]$$

with $$D_\theta = \frac{2.3}{k_\theta} \quad [4.3]$$

D_θ is the decimal reduction time, that is the time required to kill 90% of initial organisms. Where times t_1 and t_2 are necessary to obtain population N_1 and N_2, respectively (with $N_2 = \frac{N_1}{10}$) starting with population N_0, [4.2] gives indeed the following:

$$\begin{cases} \log\left(\frac{N_0}{N_1}\right) = \frac{t_1}{D_\theta} \\ \log\left(\frac{N_0}{N_2}\right) = \frac{t_2}{D_\theta} \end{cases}$$

By replacing N_1 and N_2 with their value and subtracting the two equations term by term, we obtain $D_\theta = t_2 - t_1$, corresponding to the previous definition. D_θ depends on the microorganism, its physiological state, temperature and the medium in which it is located.

The graphical representation of [4.2] in semi-logarithmic coordinates is a straight line, the inverse of the negative slope of which is D_θ (Figure 4.1). For example, a spore-forming microorganism with an initial concentration of $N_0 = 3\times10^3$ spores g^{-1} is treated at a temperature of 110°C for 10 minutes. Following this treatment, N = 3 spores g^{-1}. The change in spore number as a function of treatment time at 110°C is represented by a straight line through these two points, and D_{110} can be read directly: this is the time required to go from 10^2 to 10 spores g^{-1}, or 200 s in this case.

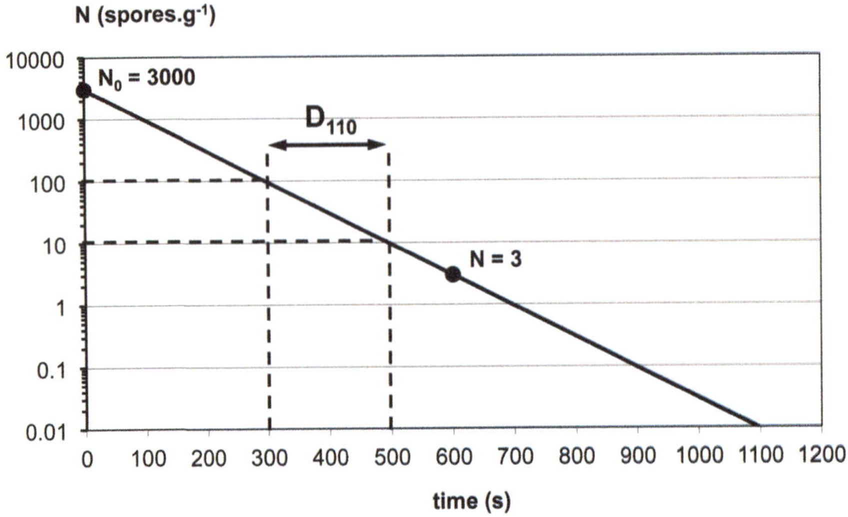

Figure 4.1. *Microbial death rate at a constant temperature*

Equation [4.2] can also be written as:

$$\begin{cases} \log\left(\dfrac{N_0}{N}\right) = n \\ t_\theta = n.D_\theta \end{cases} \qquad [4.4]$$

n is the number of decimal reductions, obtained by maintaining the product at temperature θ for time t_θ. In the previous example, $n = 3$.

Figure 4.1 shows that it is theoretically impossible to kill the entire population. Increasing the treatment time does not achieve zero probability of organisms surviving. In the previous example, it is possible to go from a 3 to 6 log reduction, or 0.003 surviving spores per g (i.e. 3 surviving spores per kg) by doubling the treatment time at 110°C (20 minutes). Although classical microbial enumeration techniques cannot ensure that microbial death always follows first-order kinetics at more intense treatment values, it is likely in practice that the survival curve drops with the length of treatment beyond a certain limit.

The parameters set by the operator are temperature θ and treatment time t_θ (which depends on the residence time in the system). Under these conditions, it is evident from [4.2] that the number of surviving microorganisms is directly dependent on N_0: in other words, the final quality of the product depends on its initial quality.

4.1.1.2. *Effect of temperature*

Increasing the treatment temperature results in an increase in the decimal reduction time; the relationship between cell death rate k_θ and temperature is expressed in the Arrhenius [2.1]. Where two temperatures θ_1 and θ_2 are in the lethal temperature range for the given microorganism and T_1 and T_2 are their associated absolute temperatures (T = 273.13 + θ), it is possible using the Arrhenius equation to establish a relationship between the associated rate constants and the temperature difference:

$$\log\left(\frac{k_{\theta 2}}{k_{\theta 1}}\right) = \frac{E}{2.3R}\left(\frac{\theta_2 - \theta_1}{T_1 T_2}\right) \qquad [4.5]$$

In the lethal temperature range, which extends from 60 to 100°C for vegetative cells and 100 to 150°C for spores, and given the values of E and R, the factor $\frac{E}{2.3R}\frac{1}{T_1 T_2}$ can be considered as a constant. For example, the activation energy for spore death is generally close to 300 kJ mol^{-1}. Between 100 and 150°C, the above expression varies from 0.11 to 0.09 K^{-1}, or an average value of 0.10 K^{-1}. The previous relationship, revised according to [4.3], is therefore:

$$\log\left(\frac{D_{\theta 1}}{D_{\theta 2}}\right) = \frac{\theta_2 - \theta_1}{z} \qquad [4.6]$$

with:

$$z = \frac{1}{\frac{E}{2.3R}\left(\frac{1}{T_1 T_2}\right)} \qquad [4.7]$$

If we assume $D_{\theta 2} = \frac{D_{\theta 1}}{10}$, then [4.6] leads to $z = \theta_2 - \theta_1$: z is the temperature increase required to reduce the decimal reduction time by one order of magnitude. In the previous example regarding spore death, z, calculated from equation [4.7], is an average value of 10°C over the temperature range of 100–150°C. This value more generally corresponds to spores, ranging from 7 to 13°C depending on the species. Vegetative cells usually have z values of between 5 and 7°C.

One aim of stabilization by heat involves carrying out *n* decimal reductions of the most resistant microorganism in the product flora. Equation [4.4] gives the corresponding treatment time t_θ at temperature θ, so equation [4.6] is therefore:

$$\log\left(\frac{nD_{\theta 1}}{nD_{\theta 2}}\right) = \log\left(\frac{t_{\theta 1}}{t_{\theta 2}}\right) = \frac{\theta_2 - \theta_1}{z} \quad [4.8]$$

The thermal resistance of spores has been studied mostly at 121.1°C (250°F), which is the reference temperature (θ_{ref}). The treatment time at this temperature ($t_{\theta ref}$) is the sterilizing value, generally denoted by $F^z_{\theta ref}$:

$$F^z_{\theta ref} = nD_{\theta ref} \quad [4.9]$$

Based on the revised expression [4.8], it is therefore possible to calculate t_θ, the equivalent time needed at any temperature θ to achieve the same number of decimal reductions than the reference treatment ($t_{\theta ref}$; θ_{ref}):

$$t_\theta = t_{\theta ref} 10^{\frac{\theta_{ref} - \theta}{z}} \quad [4.10]$$

In the same way, the concept of pasteurization unit (PU) is often used in industry, which is a conventional value corresponding to the efficiency of a heat treatment (decimal reductions) of a specific length of time and temperature (generally close to 70°C). Using [4.10], it is possible, knowing z, to calculate the number of pasteurization units of any treatment: for example, if the pasteurization unit is defined as a treatment of 1 minute at

65°C, and z is 7°C, a treatment of 1 minute at 72°C is equivalent to a treatment of 10 minutes at 65°C, or 10 PU:

$$t_{65} = t_{72} 10^{\frac{72-65}{7}} = 1(10^1) = 10 \text{ min} = 10 \text{ PU}$$

The graphical representation of [4.10] in a semi-logarithmic scale is a straight line (Figure 4.2), the inverse of the negative slope of which is z. For example, a treatment intended to achieve 6 decimal reductions of *Clostridium sporogenes* ($D_{121.1} = 57$ s), or $F_{121.1}^{10} = 6 \times 57 = 342\,\text{s}$, can be obtained by maintaining the product 4.4×10^4 s at 100°C or 44 s at 130°C. The z value can be read directly from the graph, that is the temperature increase needed to go from a treatment time of 4×10^3 to 4×10^2 s, or 10°C in this case.

It is important to note that a heat treatment must be defined as a time-temperature relationship, and that an infinite number of relationships can obtain the desired result.

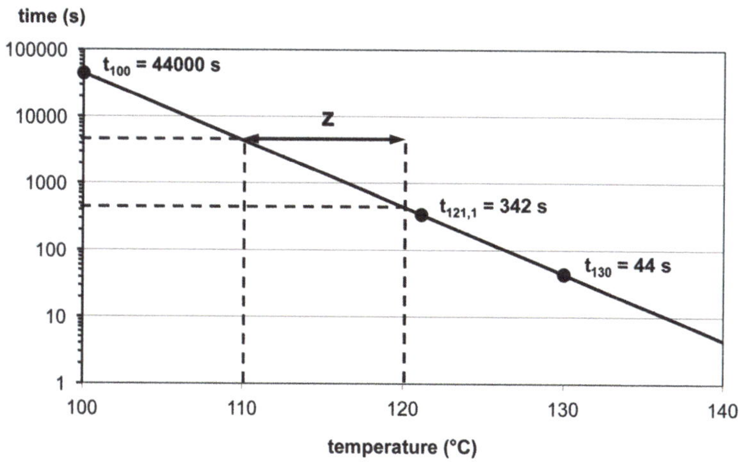

Figure 4.2. *Microbial death rate as a function of temperature*

4.1.1.3. *Effect of food properties on D_θ and z*

The physicochemical characteristics of food (pH, a_w, composition) affect the heat resistance of microorganisms. Of these characteristics, pH is by far the most influential parameter. Most vegetative or spore forms exhibit the highest heat resistance at pH close to neutrality (Figure 4.3). It can be assumed that this effect is due to a higher heat stability of intracellular enzymes at their optimal pH activity range and bacterial macromolecules (cytoplasm and wall proteins) around their isoelectric point.

Apart from heat treatment, pH has a crucial influence on the ability of bacteria to grow, and in some cases to produce toxins. It is generally assumed that the germination and growth of *Clostridium botulinum*, optimal at neutral pH, are impossible below pH 4.5. The pH of food products therefore directly determines the parameters of treatments. In this respect, there are two categories of products (Figure 4.8):

– products with a pH below 4.5: treatment at temperatures at or below 100°C can stabilize these products, with regard to both vegetative and spore forms. This rule applies only if these pH conditions are maintained in the whole product volume and for the entire product shelf life. This temperature range often corresponds to that of pasteurization, and these treatments are referred to as *sterilization*;

– products with a pH above 4.5: spore death occurs at temperatures above 100°C, generally between 115 and 150°C.

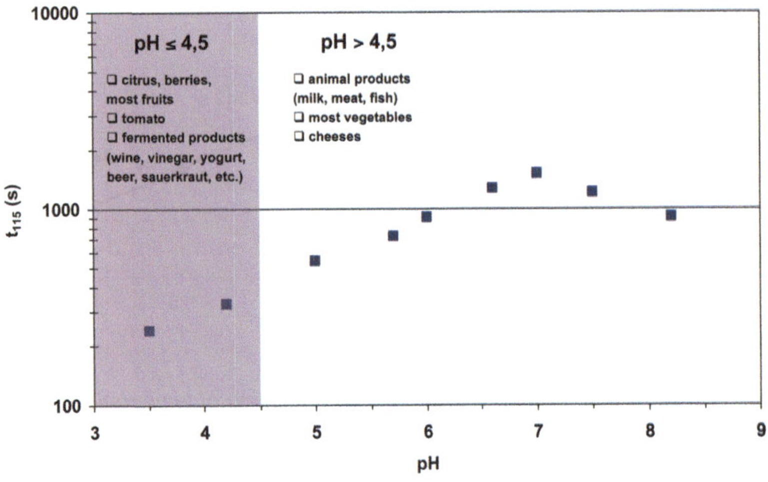

Figure 4.3. *Effect of pH on the treatment time at 115°C to obtain the same number of decimal reductions of Clostridium sporogenes*

These elements can be used to limit the value of the heat treatment and its impact on the sensory (especially textural) properties of the product. In many cases, the pH of food products is lowered by adding an acidic liquid (mackerel in white wine, addition of vinegar or organic acids to certain preserved vegetables such as pickles, etc.).

Apart from pH, a_w also affects D_θ, which is generally minimal at values close to 1, and increases when a_w decreases. Bimbinet and Loncin [BIM 95] stated that the heat resistance of spore forms was highest at a_w of about 0.3. This is due to the increase in interactions between bacterial macromolecules and other solutes at low a_w, to the detriment of those initially established with water. These interactions stabilize the native structure of these macromolecules, and require a greater amount of energy to denature them, thereby achieving microbial inactivation.

Finally, the value of D_θ can be substantially altered by the physicochemical composition of the product. For example, the presence of alkaline earth metals (Ca^{2+}, Mg^{2+}) or fats generally increases this value.

It is clear from these points that the values of D_θ given in the literature should be regarded with caution depending on the [microorganism/product] relationship. Nevertheless, Tables 4.1 and 4.2 show the ranges of D_θ and z for certain spore and vegetative forms of microorganisms.

	$D_{121.1}$ (s)	z (°C)
Bacillus cereus	2.3 – 3.9	9.7
Bacillus stearothermophilus	210 – 410	7 – 13
Clostridium botulinum	6 –18	8 – 12
Clostridium perfringens	1.3 – 1.5	7.2
Clostridium sporogenes	9 – 156	9 – 13

Table 4.1. *Values of $D_{121.1}$ and z for some spore forms of microorganisms ([BRA 77, HAL 88])*

	θ	$D_θ$ (s)	z (°C)
Non spore-forming bacteria in milk	65	35	5 – 6
Campylobacter jejuni	58.3	12 – 21	6.0 – 6.4
Enterobacter aerogenes	61.8	0.45	2.8
Enterobacter liquefaciens	73.9	0.56	6.9
Listeria monocytogenes	60	170	5.8 – 6.3
Pseudomonas pseudomonallei	73.3	0.52	9.3
Pseudomonas caryophylli	73.7	0.44	8.7
Salmonella[1]	60	12 – 390	4 – 5.6
Staphylococcus aureus	60	26 – 474	5.8 – 10
Yersinia enterocolitica	62.8	14 – 58	5.1 – 5.8

1 General data about the species; some genera are 10 to 20 times more resistant such as *Salmonella senftenberg* (z of 5.6 to 10.5°C) with a D_{60} of 570 s in eggs.

Table 4.2. *Values of $D_θ$ and z for some vegetative forms of microorganisms [BLA 82, BRA 85, KOI 83, KES 86, LOV 82]*

4.1.2. *Structural changes in food at constant temperature*

All heat treatments result in changes to the structure of food components, especially proteins (enzymes and globular proteins), vitamins (vitamin B_1) and sugars (degradation of lactose to formic acid during the Maillard reaction). Since enzymes can induce changes in food during storage, it is often necessary to inactivate them in order to ensure product stability (conformational changes may be enough to inactivate them). Finally, destroying bacterial toxins is essential to ensure food safety.

Modification of components generally follows first-order kinetics:

$$\frac{dC}{dt} = -k_{\theta}(C) \qquad [4.11]$$

where C is the concentration of the component in its native state. It should be noted that this expression has the same form as [4.1]. D_θ, in this case, is the decimal alteration time, that is to say the time required to alter 90% of the native component; on the other hand, z is the temperature increase required to reduce the decimal alteration time by one order of magnitude.

$$\begin{cases} \log\left(\frac{C_0}{C}\right) = \frac{t}{D_\theta} \\ \log\left(\frac{t_\theta}{t_{\theta ref}}\right) = \frac{\theta_{ref} - \theta}{z} \end{cases} \qquad [4.12]$$

However, second-order kinetics also exists, for example in the case of bimolecular reactions (Maillard reaction, protein aggregation).

As with microorganisms, the graphical representation of the time required to change the structure of the component to a given level as a function of time is a straight line in semi-logarithmic coordinates.

Sometimes Q_{10} is used, which is the ratio of rate constants k_θ for temperature differences of 10°C:

$$\log Q_{10} = \log\frac{k_{\theta_2}}{k_{\theta_1}} = \frac{\theta_2 - \theta_1}{z} = \frac{10}{z} \qquad [4.13]$$

For biochemical reactions with a z value of 30°C, the value of Q_{10} is 2, that is the reaction rate is doubled by increasing the temperature by 10°C.

Figure 4.4 shows the destruction of *Clostridium sporogenes* ($D_{121.1}$ = 57 s, z = 10°C) and thiamine ($D_{121.1}$ = 6,200 s, z = 25°C) in whole milk. There is a similarity between the destruction of microorganisms (i.e. a given number of decimal reductions) and change in a component (i.e. % change). However, given the differences in z, an increase of 30°C in the treatment temperature results in an increase in the rate constants (inverse of t_θ) of the reactions implied in the alteration of components ($z \cong 30$°C), the destruction of spore forms ($z \cong 10$°C) and the destruction of vegetative forms ($z \cong 5$°C) of 10, 10^3 and 10^6, respectively.

This observation forms the basis of optimizing heat treatments in the food industry. A "long time/low temperature" treatment changes the product components more than a "short time/high temperature" treatment, for the same pasteurizing or sterilizing value. Figure 4.4 shows, for example, that a decimal reduction of *Clostridium sporogenes* can be obtained in 9×10^2 s at 109°C (point A) or in 15 s at 127°C (point B). However, the first treatment leads to a 10% loss of thiamine, whereas the second treatment only alters 1%.

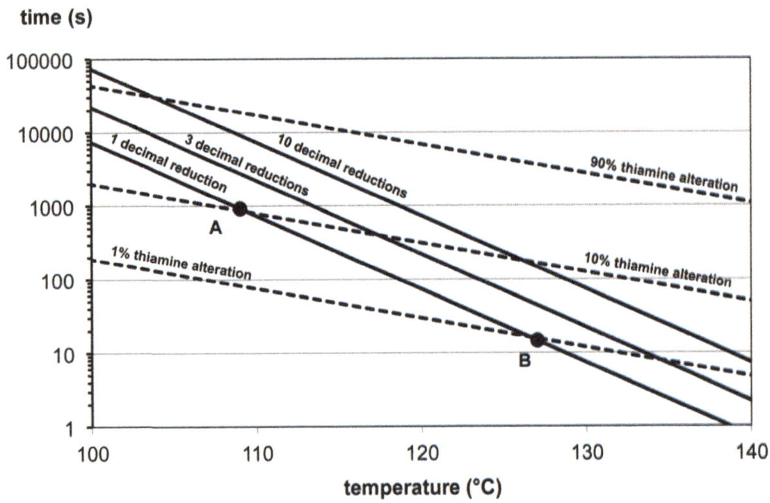

Figure 4.4. *Destruction of C. sporogenes (solid lines) and change in thiamine (dotted lines) as a function of temperature*

Tables 4.3 and 4.4 show the ranges of D_θ and z for certain constituents, bacterial toxins and global reactions taking place at temperature respectively higher and lower than 100°C.

	$D_{121.1}$ (s)	z (°C)
Vitamins		
In general	$6 \times 10^3 - 6 \times 10^4$	20 – 30
Vitamin B_1 (thiamine)	$2.3 \times 10^3 - 2.3 \times 10^4$	20 – 30
Vitamin C (ascorbic acid)	1.5×10^5	51
Pantothenic acid	$1.5 \times 10^5 - 3.8 \times 10^6$	31
Vitamin B_2 (riboflavin)	1.7×10^6	28
Folic acid	1.7×10^6	37
Enzymes		
In general	60 – 600	7 – 55
Lipase (from *Pseudomonas*)	616 – 872	25 – 37
Protease (from *Pseudomonas*)	240 – 816	32
Other transformations		
Maillard reaction coloration	$24 - 2.4 \times 10^3$	17 – 39
Loss of lysine	4.5×10^4	21

Table 4.3. *Values of $D_{121.1}$ and z for certain components and reactions at temperatures above 100°C [HAL 88]*

Some enzymes and bacterial toxins, such as proteases and lipases from *Pseudomonas* or the toxin from *Staphylococcus aureus*, are very heat stable (more than those of the given microorganisms), therefore highlighting again the influence of the initial quality of the product on its final quality.

	θ	D_θ (s)	z (°C)
Alkaline phosphatase	72	8	5
Milk lipase	72	5.9	6.4
Peroxidase	72	1.6×10^3	9
Catalase	72	142	7.2
Serum proteins	72	8.1×10^3	6
Bacterial toxins			
Staphylococcus aureus	98.9	$> 7.2 \times 10^3$	27.8
Clostridium botulinum	85	3×10^2	4.0 – 6.2

Table 4.4. *Values of D_θ and z of denaturation reactions of certain components and bacterial toxins [DEN 71, KES 86, REA 66, WOO 79]*

4.1.3. *Heat treatment at variable temperature*

In most cases, the temperature of the treated product changes over time during heat treatment. The plateau phase at which the product is maintained at the target operation temperature (θ_p) is preceded by a heating phase and followed by the cooling phase.

This temperature change can be divided into a series of basic processing units with a time of Δt, which are short enough so that the temperature can be considered constant and equal to θ_i during this interval. It is therefore possible to calculate a decimal reduction time $D_{\theta i}$ for each interval, resulting in:

$$\begin{cases} \log\left(\dfrac{N_0}{N_1}\right) = \dfrac{\Delta t}{D_{\theta 1}} \\ \log\left(\dfrac{N_1}{N_2}\right) = \dfrac{\Delta t}{D_{\theta 2}} \\ \cdots\cdot \\ \log\left(\dfrac{N_{n-1}}{N_n}\right) = \dfrac{\Delta t}{D_{\theta n}} \end{cases} \qquad [4.14]$$

The sum of the terms is:

$$\log\left(\frac{N_0}{N_n}\right)=\sum_{i=1}^{n}\frac{\Delta t}{D_{\theta i}} \qquad [4.15]$$

Given that $\log\left(\frac{N_0}{N_n}\right)$ is *n*, the number of decimal reductions, and by tending Δt to an infinitely small dt, the result is:

$$n=\frac{F^{z}_{\theta ref}}{D_{\theta ref}}=\int_0^t\frac{dt}{D_\theta} \qquad [4.16]$$

with $D_\theta = D_{\theta ref}10^{\frac{\theta_{ref}-\theta}{z}}$. Since $D_{\theta ref}$ is a constant, $F^{z}_{\theta ref}$ can be calculated using [4.9] and [4.16]:

$$F^{z}_{\theta ref}=D_{\theta ref}\int_0^t\frac{dt}{D_\theta}=\int_0^t 10^{\frac{\theta-\theta_{ref}}{z}}dt \qquad [4.17]$$

This equation can be solved by the Bigelow's method, which consists in calculating the equivalent of each time interval (short enough to be able to consider the product temperature θ constant) at the reference temperature θ_{ref}. $F^{z}_{\theta ref}$ is obtained by adding all these separate equivalent times. This calculation can easily be performed using spread-sheet programs.

By way of example, Figure 4.5 shows the change in temperature at the center of a container during the sterilization of a food product, the instantaneous values of $\frac{D_{\theta ref}}{D_\theta}$ and the cumulative value of $F^{z}_{\theta ref}$ as a function of time. The change in temperature described here is characteristic of transfer by pure conduction, corresponding to an immobile product. It can be estimated using the Newman expression (see Chapter 1, section 1.1.1.4), if the thermal diffusivity of the given food is known. In practice, the time spent at a temperature θ lower than $[\theta_p - 2z]$ (or θ = 90°C for θ_p = 110°C and z = 10°C; grey area) contributes negligibly to the cumulative sterilizing value $F^{z}_{\theta ref}$. In addition, the inclusion of any time interval in this calculation assumes that the corresponding temperature θ is deemed to have a sterilizing effect on the pH of the product.

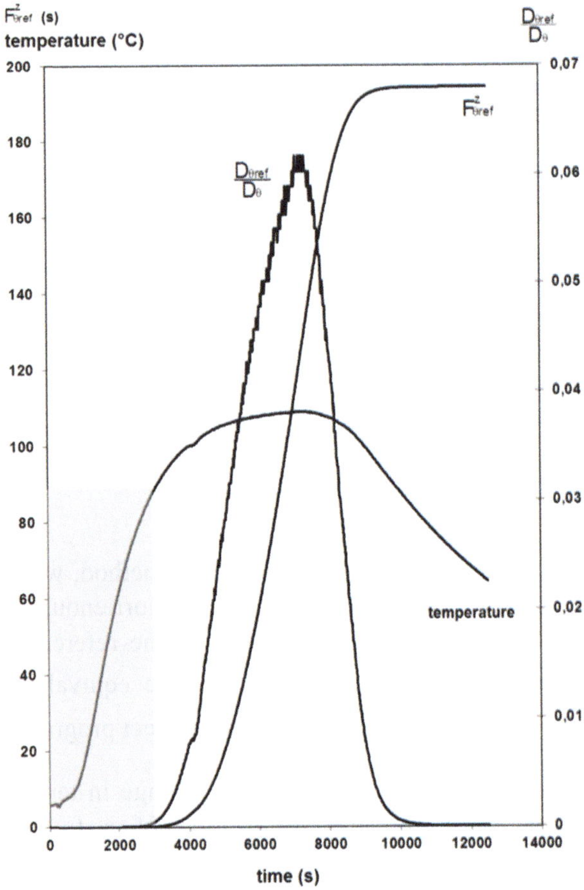

Figure 4.5. *Change in temperature and $F^{z}_{\theta ref}$ as a function of time during sterilization*

4.1.4. *Practical aspects of heat treatments*

4.1.4.1. *Target pasteurizing or sterilizing values*

Pasteurization and sterilization are used to improve hygienic quality or ensure long shelf life.

Pasteurization

In order to extend product shelf life, at least a 5 decimal reduction of vegetative cells must be obtained. Three time-temperature combinations are used: low temperature pasteurization (15 – 30 min/60 – 65°C), high temperature pasteurization (15 – 40 s/70 – 75°C) and flash pasteurization (1 – 2 s/85 – 95°C). Given the short duration of high and flash pasteurization, these methods are usually used for liquids. The treatment used to ensure the safety of the product until the use-by date remains the responsibility of the manufacturer, provided that it has been approved by the appropriate regulatory authority: thus, higher heat treatments than those previously mentioned ($\theta > 100°C$, especially ultra-high temperature treatments), followed by non-aseptic packaging, can be used while still maintaining the term "high pasteurization". The residual activity of enzymes is a good indicator of the type of heat treatment. Thus, a low-pasteurization treatment of milk should inactivate alkaline phosphatase but preserve peroxidase ("fresh pasteurized milk").

Sterilization

This treatment should at least correspond to a 12 decimal reduction of *Clostridium botulinum* ($D_{121.1°C} \cong 12$ s, where $F_{121.1}^{10} = 144$ s), or a sterilizing value usually rounded off to 180 s. Bimbenet and Loncin [BIM 95] pointed out that the obligation on the manufacturer to lower the risk of poor preservation to a commercially acceptable level by the destruction of non-pathogenic spores (e.g. at least one organism per 10^4 containers) results in much higher target values for $F_{121.1}^{10}$. Two time-temperature combination ranges are used: 15 – 20 min/115 –125°C and 2 – 6 s/140 –150°C (ultra-high temperature or UHT). UHT treatment limits the loss of essential amino acids as well as Maillard browning. The need to preserve the product constituents in a native form should not overshadow the need to destroy certain enzymes that may destabilize the product during its preservation (lipases of psychrotrophic bacteria for example).

4.1.4.2. *Bulk processing*

Bulk processing is mainly used for liquid products and can be performed either in batch or continuous mode. In this case, the product is packaged after heat treatment. This is a critical point in the process as it involves a risk of recontamination in the case of non-aseptic packaging.

Batch process

A pasteurization process can be carried out by heating the liquid in a stirred vessel (Figure 4.6).

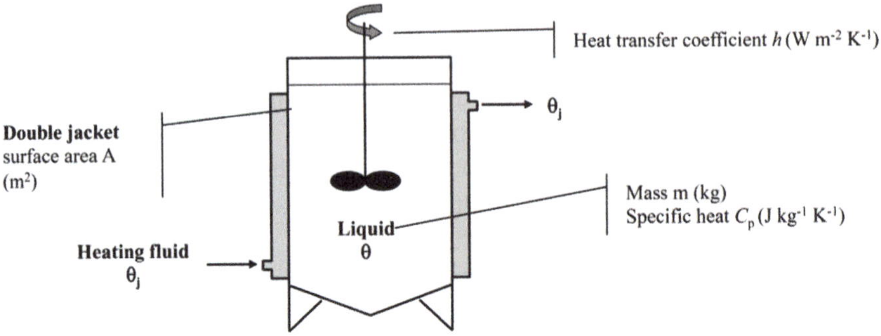

Figure 4.6. *Heating of liquid in a jacketed stirred tank*

If the temperature of hot water circulating through the jacket is θ_j (assumed constant over time) and the initial temperature and the temperature at time t of milk are θ_0 and θ respectively, we can write:

$$\mathrm{m}C_\mathrm{p}\frac{\mathrm{d}\theta}{\mathrm{dt}} = \mathrm{A}\ h\ (\theta_j - \theta) \qquad [4.18]$$

where m (kg) and C_p (J.kg^{-1}.K^{-1}) are, respectively, the mass and specific heat of the treated liquid, A is the exchange surface area (m^2), and h is the heat transfer coefficient (W.m^{-2}.K^{-1}). This expression is integrated to:

$$\log\left(\frac{\theta_\mathrm{j} - \theta}{\theta_\mathrm{j} - \theta_0}\right) = -\frac{\mathrm{A}\,h}{2.3\mathrm{m}\,C_\mathrm{p}}\mathrm{t} \qquad [4.19]$$

In a semi-logarithmic plot versus time, the term $\frac{\theta_j - \theta}{\theta_\mathrm{j} - \theta_0}$, also called reduced temperature (see Chapter 1, section 1.1.4), is linear. This equation describes as well the case of pre-packaged liquid (e.g., milk in a bottle), which temperature θ can be considered uniform over its entire volume.

If the change in temperature over time is known, $F^{z}_{\theta ref}$ can be calculated as described in section 4.1.3.

Continuous process

The continuous heat treatment of liquids (pasteurization or sterilization) can be carried out in tube or plate exchangers, connected to a holding tube to maintain the treatment temperature. In a heat exchanger the heat transfer fluid delivers or removes heat to or from the product.

Principles of heat exchangers

In tube heat exchangers, the product and the heat transfer fluid flow through concentric tubes, one in the central space and the other in the annular space. The advantage of this type of system is the high circulation speed of the product resulting in turbulent flow, which limits problems of fouling and thereby making it possible to treat viscous products. However, this type of exchanger tends to be bulky and non-versatile.

Plate heat exchangers are, in contrast, very compact and easily maintained, but the pressure drop between the plates is high; the required pump must therefore withstand high back pressure. Moreover, obstructions between the plates are possible due to the limited space and the low circulation speed of the product between the plates. These heat exchangers are therefore used for less viscous products.

For the most viscous products, heat transfer can be improved by the action of blades or knives that scrape the exchange surface facilitating the renewal of the boundary layer and the mixing of the liquid mass: this is the principle of scraped surface heat exchangers.

The general principle of continuous heat treatment is given in Figure 4.7. It involves bringing the liquid to its treatment temperature through heat exchange with a heating fluid, maintaining it at the treatment temperature θ_h for the required holding time, and then cooling it. When a fluid passes through a heat treatment device, it remains there for an average period of time defined by the ratio between the volume of the device and the volume flow of the fluid. However, some volume elements move faster than the average while others are slower; this phenomenon is known as the residence time distribution and depends on the device. The over-treatment of microorganisms that remain longer in the device does not compensate for the under-treatment of microorganisms that are there for a shorter time. In practice, this residence time distribution requires an increase in the average treatment (i.e., mean residence time) in order to reach the target pasteurizing (or sterilizing) value, and therefore has a greater impact on the product. [BAM 02] give a broad overview of how to characterize the distribution of residence times in a facility.

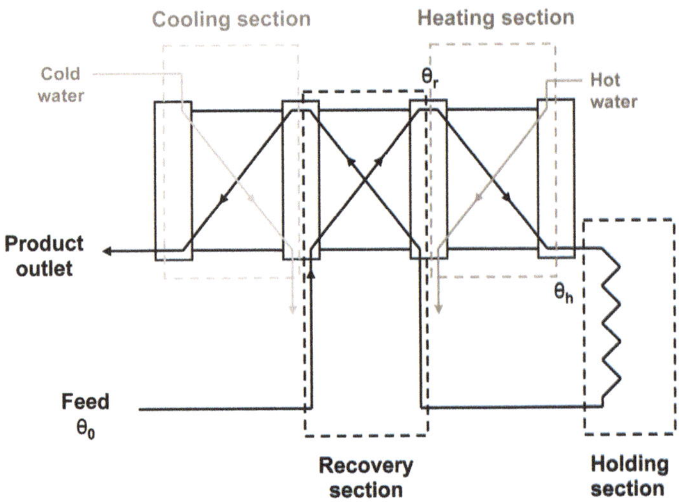

Figure 4.7. *Basic diagram of a continuous heat treatment facility*

In order to reduce energy costs, it is possible to recover heat, whereby the "rising" product (to the holding section) is heated to θ_r and the "falling" product (from the holding section) is cooled. The heating zone (by hot water or steam) is necessary to rise the temperature of the product from θ_r to θ_h, because the heat recovery coefficient r_q of the facility is usually around 90–92%. An additional cooling zone (cold water or glycol water) is generally used to bring the product to storage temperature.

Heat exchanges in each of these different zones are counter-current to allow a better recovery rate. However, co-current heat exchangers can also be used, even though in this case the temperature difference between the heat transfer fluid and the product at the inlet of the heat exchanger is very high and therefore prone to wall fouling.

Principles of direct and indirect sterilization

The sterilization of liquids can be carried out continuously at a high temperature for a short period of time (ultra-high temperature, or UHT treatment). Direct UHT treatment involves injecting live steam into a preheated liquid (temperature θ_1) in the heat exchanger. The condensation of steam in the liquid results in the release of latent heat of vaporization, which causes an almost instantaneous temperature increase (temperature θ_2) along with a dilution of the product (Figure 4.8):

$$\dot{m}_s H + \dot{m}_p C_p \theta_1 = \dot{m}_s C_{pc} \theta_2 + \dot{m}_p C_p \theta_2 \qquad [4.20]$$

where $\dot{m}_p$ and $\dot{m}_s$ are the flow rates of the product and steam (kg s^{-1}), C_p and C_{pc} are the specific heat of the product and the condensed steam respectively (J kg^{-1} K^{-1}) and H is the enthalpy of the live steam (J kg^{-1}). The diluted product is then reconcentrated in an expansion chamber through the evaporation of water supplied during heating, resulting in instant cooling:

$$\dot{m}_s C_{pc} \theta_2 + \dot{m}_p C_p \theta_2 = \dot{m}_v H' + \dot{m}_p C_p \theta' \qquad [4.21]$$

where θ' is the temperature of the cooled product and $\dot{m}_v$ (kg s^{-1}) and H' (J kg^{-1}) are the mass flow rate and enthalpy of the secondary vapor generated at this temperature, respectively. Process control therefore depends on the adjustment between the amount of steam injected during heating ($\dot{m}_s$) and the amount of secondary vapor removed by expansion during the cooling of the product ($\dot{m}_v$).. Food-grade steam must be used in direct UHT treatment. To avoid the mixing of steam with the product, indirect UHT treatment with a tube heat exchanger can be used. The temperature increase is slower in this case: since the heat shock is greater, destabilization and fouling are more likely to occur.

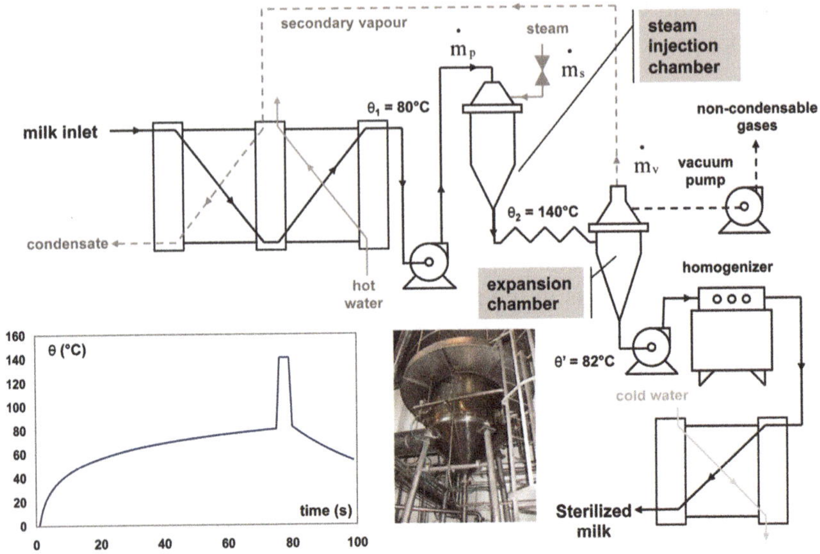

Figure 4.8. *Direct UHT milk sterilization system and corresponding time-temperature profile*

Sizing of heat exchangers

Determining the size of the different zones in a counter-current heat exchanger can be based on the energy of these zones (Figure 4.9). It involves stating that the power supplied by the hot fluid is equal to the power received by the cold fluid and the power transferred across the heat exchange surface (Fourier's law). In the example in Figure 4.9, fluid 1 is the hot fluid and fluid 2 is the cold fluid; assuming negligible heat losses, the following generally applies:

$$\dot{Q} = \dot{m}_1 C_{p1} (\theta_{1i} - \theta_{1o}) = \dot{m}_2 C_{p2} (\theta_{2i} - \theta_{2o}) \quad [4.22]$$

with $\dot{Q}$ (W) being the heat power exchanged within the unit; $\dot{m}_1$, $\dot{m}_2$ (kg s^{-1}) the mass flow rates of fluids 1 and 2; C_{p_1}, C_{p_2} (J kg^{-1} K^{-1}) the specific heat capacities of fluids 1 and 2, and θ_{1i}, θ_{2i} and θ_{1o}, θ_{2o} (°C) the temperatures of fluids 1 and 2 at the inlet and the outlet of the heat exchanger, respectively.

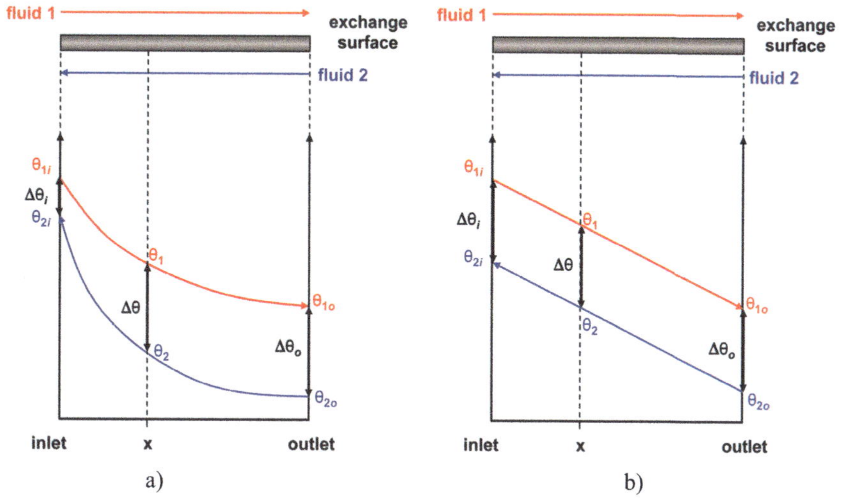

Figure 4.9. *Sizing of a counter-current heat exchanger. a) general case; b) case of recovery zones*

The flow rates are different in the heating or cooling zones using a heat transfer fluid: $\Delta\theta$ and consequently the power exchanged per unit area varies throughout the heat exchanger (Figure 4.9(a); inlet difference of temperature between the two fluids $\Delta\theta_i$ differing from the outlet difference $\Delta\theta_o$). In this case, Hausbrand's formula is used to express Fourier's law:

$$\dot{Q} = A\,h\,\frac{\Delta\theta_o - \Delta\theta_i}{\ln\dfrac{\Delta\theta_o}{\Delta\theta_i}} = A\,h\,\Delta\theta_m \qquad [4.23]$$

where A (m^2) is the exchange surface area, α (W m^{-2} K^{-1}) is the overall heat transfer coefficient and $\Delta\theta_m$ is the logarithmic mean temperature difference between fluids 1 and 2. This mean value represents the factor which potentiates the heat transfer through the exchange surface. By combining equations [4.22] and [4.23], we obtain:

$$\dot{Q} = A\;h\;\Delta\theta_m = \dot{m}_1\,C_{p1}\,(\theta_{1i} - \theta_{1o}) \qquad [4.24]$$

where:

$$A = \frac{\dot{m}\, C_{p1}}{h} \, \frac{\theta_{1i} - \theta_{1o}}{\Delta\theta_m} \qquad [4.25]$$

The term $\frac{\theta_{1i} - \theta_{1o}}{\Delta\theta_m}$ represents the efficiency factor that compares the result obtained for fluid 1 $(\theta_{1i} - \theta_{1o})$ to the mean values ($\Delta\theta_m$). This dimensionless ratio is called the number of transfer units for fluid 1, or NTU_1. The term $\frac{\dot{m}\, C_{p1}}{h}$ has the dimension of a surface and is independent of temperature. It is known as the surface transfer unit for fluid 1, or A_1. [4.25] is therefore:

$$A = A_1\, NTU_1 \qquad [4.26]$$

The same applies for fluid 2, where:

$$A = A_1\, NTU_1 = A_2\, NTU_2 \qquad [4.27]$$

The heat balance for the entire heat exchanger can be calculated based on this equation.

In the recovery zone (Figure 4.9 (b)), the mass flow rate is often the same on both sides of the exchanger. $\Delta\theta$ is therefore constant at every point in the recovery zone, giving:

$$\dot{Q} = \dot{m}\, C_p\, (\theta_{1i} - \theta_{1o}) = A\, h\, \Delta\theta \qquad [4.28]$$

If the heat transfer coefficient h and the inlet and outlet temperatures of the fluid are known, it is therefore possible to calculate the temperature difference between the two fluids ($\Delta\theta_m$ or $\Delta\theta$ depending on the case) and consequently the exchange surface A of the heat exchanger. Table 4.5 gives values of h in common cases.

Fluid	*h* $(W\,m^{-2}\,K^{-1})$
Still air	3 to 23 (depending on H_r)
Moving air	10 to 10^2
Air fluidized bed	30 to 60
Still water	3×10^2
Stirred water	10^3 to 2.3×10^3
Milk in a pasteurizer	1.2×10^3 to 2×10^3
Brine	9×10^2
Condensing steam	6×10^3 to 6×10^4
Steam sterilization	1.7×10^4
Steam sterilization + 50% air	1.7×10^3

Table 4.5. *Some values of h in common heat exchange conditions*

However, in most cases these values are not precise enough and a more detailed evaluation of *h* must be carried out (see Chapter 1, section 1.2.5.4). In the case of plate heat exchangers, a process relationship between dimensionless number applicable in the turbulent regime (Re > 1,000) makes it possible to obtain the *h* value:

$$\mathrm{Nu} = 0.218\,\mathrm{Re}^{0.67}\,\mathrm{Pr}^{0.4} \qquad [4.29]$$

The hydrodynamic conditions should be such that flow is turbulent to avoid fouling while controlling the residence time distribution. The turbulent flow conditions are obtained either by increasing the flow or using non-planar geometries (turbulence promoters).

The dimensions of a heat treatment facility are a compromise between:

– the heat recovery coefficient r_q, an increase in which results in a lowering of energy costs and an increase in the exchange surface and therefore depreciation costs. This parameter directly depends on the energy cost and the unit cost of the exchange surface;

– the thermal sensitivity of the product, which determines the temperature difference between the hot and cold fluid ($\Delta\theta_m$ or $\Delta\theta$ depending on the case);

– the pressure drop in the facility, which can be a technical limitation or a prohibitive operating cost;

– the choice of plate sizes and geometries, which is linked to pressure drop and heat exchange coefficients values.

In conclusion, the heat exchanger should represent a compromise between high h values and acceptable pressure drop.

4.1.4.3. *Post-packaging treatment*

After packaging, most food, whether solid, liquid or mixed can be heat-treated in autoclaves either batchwise or continuously. Unlike the previous case, the risk of recontamination of the product is very limited (damage to container during treatment or faulty sealing) since the packaging is carried out before the heat treatment.

Depending on the type of food, heat transfer is either by conduction (solids) or convection (liquids) (Figure 4.10). The transfer rate, even for liquids, remains well below those possible for bulk treatment due to low heat transfer coefficients h. Rotating the container and stirring the heat transfer medium, while reducing the boundary layers, improves heat transfer (increase in h) and reduces product defects after heat treatment. In all cases, the sterilizing value $F^{z}_{\theta ref}$ in the geometric centre of the container can be determined based on the recorded temperature profile.

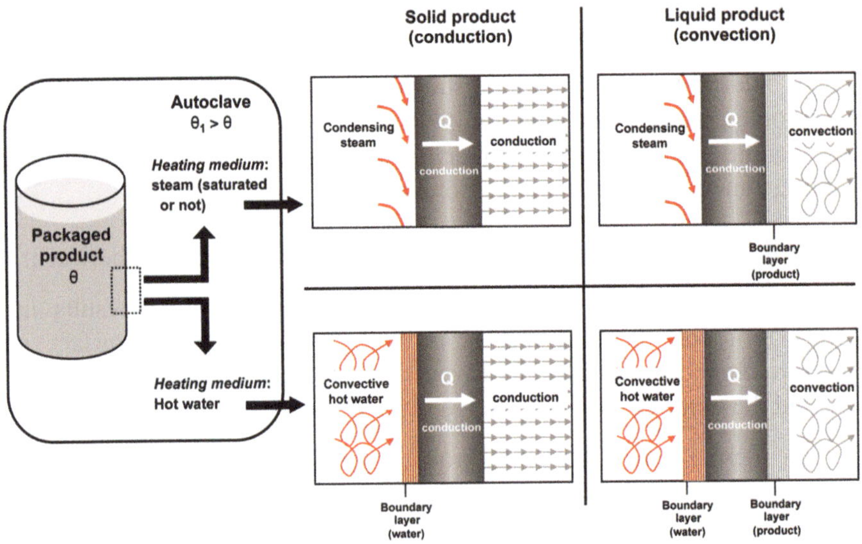

Figure 4.10. *Heating of the product after packaging*

Heating a product in a hermetically-sealed container causes an increase in its internal pressure. This is a limiting factor in terms of the maximum operating temperature, which is generally less than or equal to 130°C (corresponding to a maximum operating pressure of around 0.3 MPa).

Thus, treatment after packaging means longer heating times at lower temperatures compared to time-temperature combinations used in bulk treatments.

Packaging: calculating the headspace

The filling of containers begins, where applicable, with the larger solid elements and ends with the addition of the liquid medium (oil, brine, sauce). Pre-treating solid portions facilitates handling (hardening of sardines in brine) and limits exudation and shrinkage after heat treatment (tumbling of meat).

In order to avoid an increase in pressure inside the packaging during the heating phase, it is necessary to leave a headspace (dead volume of air) during filling. This is particularly important when the thermal expansion of the product is greater than that of the container (glass, metal). It is possible to determine the limit value of this headspace (proportion α (%) of the packaging volume V_0) based on the expansion coefficients of the packaging and the product as well as the temperature variation during treatment (Figure 4.11). At the initial temperature θ_0, the volumes of the packaging, the headspace and the product are, respectively, V_0, αV_0 and $(1-\alpha)V_0$. If β_p and β_e are, respectively, the volume expansion coefficients of the product and the packaging ($m^3 K^{-1}$), the volume increase of the container at temperature θ is:

$$\beta_c \, V_0 \, (\theta - \theta_0) \tag{4.30}$$

At this temperature, the volume increase of the product is:

$$\beta_p \, (1-\alpha) V_0 \, (\theta - \theta_0) \tag{4.31}$$

In order to prevent excess pressure, the difference between the volume increase of the product and that of the packaging must be less than or equal to the headspace volume (αV_0). The limit condition is therefore:

$$\beta_p \, (1-\alpha) V_0 \, (\theta - \theta_0) - \beta_c \, V_0 \, (\theta - \theta_0) \leq \alpha \, V_0 \tag{4.32}$$

Hence:

$$\alpha \geq \frac{(\beta_p - \beta_c)\ (\theta - \theta_0)}{1 + \beta_p\ (\theta - \theta_0)} \qquad [4.33]$$

The volume expansion coefficients β_c are, respectively, 1.8×10^{-5}, 3.6×10^{-5} and 6×10^{-4} m^3 °C^{-1} for glass, steel and polyethylene. For a liquid product ($\beta_p = 4.5\times10^{-4}$ m^3 °C^{-1}; θ_0 = 50°C; θ = 120°C), the limit value below the headspace calculated by [4.33] is 3% for a glass container and 2.8% for a steel container. This calculation only applies where $\beta_p > \beta_c$. Otherwise it gives negative values (polyethylene packaging).

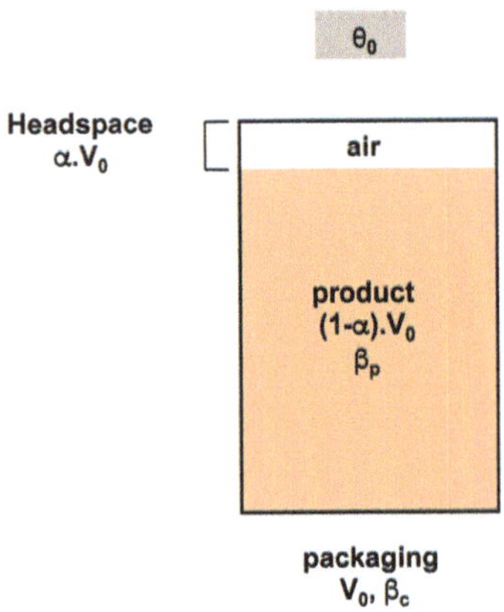

Figure 4.11. *Calculation of the headspace value*

Temperature and pressure changes

A temperature change during the heating and cooling phases results in a pressure change in the autoclave and the product. A direct consequence of these pressure and temperature variations is the mandatory installation of airlocks at the inlet, outlet and between the different zones (heating, cooling) of continuous autoclaves.

During heating, the temperature θ of the product is less than that of the ambient air in the autoclave, and the pressure difference [autoclave – product] is positive. The main risk would be a sudden rise in temperature in the autoclave, which could, in extreme cases, crush the container.

In contrast, the pressure difference [autoclave – product] is negative (excess pressure inside the container compared to the heating/cooling medium in the autoclave):

– during the plateau phase at target operation temperature if the product contains air: in this case, even though the temperature of the heating medium and the product are almost identical, the presence of air in this latter results in a higher internal pressure in the packaging compared to that of the autoclave;

– during the cooling phase: the temperature and pressure of cooling medium are lower than those of the product, especially at the beginning of cooling.

The risks associated with this scenario are recontamination of the product after heat treatment due to defective sealing, and in extreme cases, a bursting of the container due to pressure differences.

It is therefore essential to correctly manage the pressure difference between the autoclave and the product. The risk of crushing may be limited by controlling and extending the heating time in the autoclave. The risk of bursting can be decreased on the whole by ensuring sufficient headspace and filling the container under vacuum. It is also possible to vary the pressure in the autoclave by injecting compressed air into it during cooling so as to limit and even eliminate the pressure difference with the product.

One original technical solution to overcome this pressure differential during the heating and cooling phases is known as a "hydrostatic" autoclave system (Figure 4.12): the product enters and leaves the sterilization zone via two vertical water columns; the height of the columns gradually compensates for the movement of the containers, the pressure increase during heating and the pressure decrease during cooling.

4.2. Food irradiation

Food irradiation is a process of exposing food to ionizing radiation to sterilize it and/or increase its shelf life by destroying microbial activity. Despite being developed about 30 years ago, its applications are limited: this process is not intended to replace stabilization treatments based on thermal destruction or the reduction of a_w, but is used for raw or heat-sensitive products. In addition to destroying microbial flora, it also kills insects responsible for crop losses and halts physiological processes such as the germination of plant tubers. One of the advantages of this method is that is can be used on frozen and pre-packaged products.

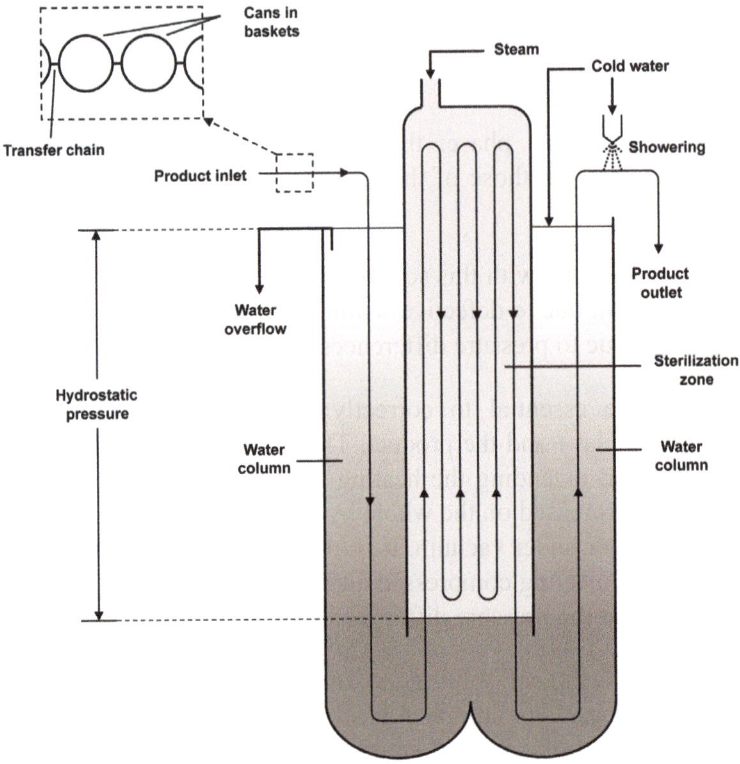

Figure 4.12. *Diagram of a continuous hydrostatic sterilizer*

4.2.1. *Principle*

Irradiation treatment involves subjecting food to electromagnetic γ radiation from the radioactive decay of cobalt 60 or caesium 137, from X-rays of energy below 5 MeV or an accelerated electron beam of less than 10 MeV (1 Mev = 1.6 10^{-13} J).

$$ {}^{60}_{27}\text{Co} \rightarrow {}^{60}_{28}\text{Ni} + e^{-} + \gamma \quad (1.20\text{ Mev}) \qquad [4.34] $$

$$^{137}_{55}Cs \rightarrow {}^{137}_{56}Ba + e^{-} + \gamma \quad (0.66\ \text{Mev}) \qquad [4.35]$$

These treatments are referred to as 'irradiation' since radioactive sources are used, which might suggest that they generate radioactivity. In reality, and given the energy levels used, this is not the case.

Electromagnetic γ radiation and electron beams disturb the electronic energy levels of atoms, especially those with high electron density (oxygen, nitrogen, sulfur). This can result in the breaking of covalent bonds and the formation of free radicals or ionizations by the removal of an electron. The elements formed by radiolysis are highly reactive and can continue to be active after treatment and during storage. Water from tissues and foods gives free radicals and ions as follows:

$$\begin{aligned} H_2O &\rightarrow HO^{\bullet} + H^{\bullet} \\ H_2O &\rightarrow H_2O^{+} + e^{-} \end{aligned} \qquad [4.36]$$

In the presence of oxygen, hydrogen peroxide can form:

$$\begin{aligned} H^{\bullet} + O_2 &\rightarrow HO_2^{\bullet} \\ HO_2^{\bullet} + HO_2^{\bullet} &\rightarrow H_2O_2 + O_2 \end{aligned} \qquad [4.37]$$

Cell components, including DNA, undergo changes either as a direct result of radiation or electrons, or by interaction with products from radiolysis. These structural changes to DNA lead to the destruction of microorganisms. Treatment can also cause a loss of nutrients, including certain vitamins, and generate volatile products that affect the sensory quality of the food through the degradation of sulfur amino acids and lipids. To minimize these effects, irradiation is sometimes carried out in the frozen state, which slows down the transfer of reaction products in the food.

Fatty acid molecules such as cyclobutanones may also form by cyclization, which are potentially toxic:

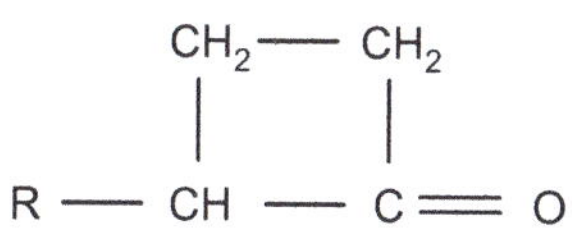

[4.38]

The effect of irradiation treatment on the inactivation of microbial enzymes and toxins is relatively limited. However, the objective of stabilization treatments is to inhibit bacterial growth and enzymatic reactions while at the same time destroying pathogens and toxins that have been released into the food. In this respect, stabilization by irradiation is not as effective as by heat treatment.

The positive and negative effects of the treatment depend on the "dose" applied: it is measured in Gray (Gy), the unit energy corresponding to the absorption of one joule per kg of product (J kg^{-1}). The dose depends on the type and energy of the radiation (γ rays are more penetrating than electrons), the exposure time and the shape of the product. The effectiveness of the treatment decreases with the thickness of the product since electromagnetic radiation and electrons are slowed down by their interaction with matter: the effect of the treatment is therefore more pronounced on the surface than on the insidc.

The intensity of the treatment depends on the purpose:

– Inhibition of germination: 0.04 to 0.10 kGy;

– Destruction of insects: 0.10 to 3 kGy;

– Partial destruction of microorganisms (*radicidation*, equivalent to heat treatment): 1 to 4 kGy;

– Destruction of pathogenic bacteria (*radurization*, equivalent to pasteurization): 1 to 10 kGy;

– Complete destruction of microorganisms (*radappertization*, equivalent to sterilization): 15 to 50 kGy.

The energy levels used are still very limited as they do not exceed 50 $kJ.kg^{-1}$.

4.2.2. Destruction of microorganisms

The destruction of microorganisms by irradiation treatment can be defined by a relation similar to that of heat treatments:

$$\frac{dN}{dE} = -k\,N \qquad [4.39]$$

where N is the number of microorganisms at time t, E is the energy used (Gy) and k is the destruction constant (Gy^{-1}) dependent on microorganisms. This equation is integrated between a population N_0 at time 0 and a population N at time t in:

$$\log\left(\frac{N_0}{N}\right)=\frac{E}{E_{10}} \quad [4.40]$$

with $$E_{10}=\frac{2.3}{k} \quad [4.41]$$

E_{10} is the radiation decimal reduction dose, that is the energy required to destroy 90% of the initial flora. For example, E_{10} is equal to 0.06 kGy for *Pseudomonas,* 0.70 kGy for *Salmonella,* 4 kGy for *Clostridium botulinum* and 14 kGy for the poliovirus.

A treatment of 4 kGy thus obtains about 6 decimal reductions of *salmonella* and about 10 decimal reductions of spoilage flora. This treatment sterilizes the product and prolongs shelf life in cold storage in the same way as pasteurization. A dose of 48 kGy must be applied to obtain the equivalent of heat sterilization, which aims to achieve 12 decimal reductions of *Clostridium botulinum.* However, this treatment does not have the same effect on the destruction of enzymes, viruses and toxins, and does not ensure the same level of product stabilization.

Food irradiation can be performed at low temperatures and also after packaging. These characteristics make it an effective technology to deal with raw and heat-sensitive products, for which heat treatment is not suitable, and to handle products that may have been contaminated during packaging.

4.2.3. *Areas of application*

As the safety of irradiated food has been recognized by the World Health Organization (WHO) for doses below 10 kGy since 1980 and for all doses since 1999, the process has been authorized in many countries but still remains banned in others. The WHO supports the development of this process especially given its increasing concern about food contamination.

Despite this position, many consumer groups in the U.S. and Europe are opposed to such technology because they believe it poses a risk to the environment, operators and consumers. The FDA (Food and Drug Administration) has however authorized the irradiation of fresh or frozen beef, lamb and pork in 1997 in order to eliminate bacteria such as E. coli. The situation in Europe is more complex since some countries such as Germany are against it while others like France, Belgium and the Netherlands are very much in favor of it. The list of foods authorized for irradiation in Europe is limited to three product categories:

– Dried aromatic herbs;

– Spices;

– Vegetable condiments.

The products involved in the preparation of ready meals are sources of contamination that could affect product quality during storage.

Given the difficulty in reaching a consensus between member states, European legislation continues to recognize national authorization. The European Parliament remains very cautious about extending the list of products for which irradiation is authorized due to the potential health risks for consumers and employees as well as possible environmental pollution.

In 2001, the French agency for food safety (*Agence Française de Sécurité Saritaire des Ailments* (AFSSA)), approved the irradiation of food intended for human and animal consumption, as it considers the process safe for the consumer. However, it states that this process can only be applied to foods produced using good manufacturing and hygiene practices. Authorization covers:

– spices and flavorings (decontamination: 10 kGy);

– onions, garlic and shallots (anti-germination: 0.075 kGy);

– dried fruit and vegetables (elimination of insects: 1 kGy);

– poultry meat, mechanically separated meat, offal, frozen frogs legs, frozen prawns, egg whites (destruction of pathogens: 3 to 5 kGy);

– dried blood, plasma, caseinates (bacterial decontamination: 6 to 10 kGy).

At present, the amount of foodstuffs treated by irradiation still remains relatively low: less than 20,000 tonnes in Europe in 2002. However, irradiation methods are currently used in packaging and the treatment of medical equipment (sterilization of syringes, surgical gloves, compresses, implants and prosthetics).

4.2.4. *Detection of irradiated food*

Consumers themselves have no way of identifying irradiated foods. In Europe, all irradiated foods must be labeled "treated with ionizing radiation" or "treated by irradiation". There are currently a number of methods for identifying irradiated foods. In the case of raw foods, the level of damage to microbial DNA is a good indicator; however in cooked products, it is difficult to distinguish between thermal or irradiation damage.

Electron spin resonance (ESR), thermoluminescence and gas chromatography techniques can also be used to determine the presence of substances produced by food irradiation (free radicals, peroxides, cyclobutanones, etc.). However, it is important to remember that some of these components can also be formed during heat treatment. The analysis of cyclobutanones would probably be the most sensitive and the most relevant for products containing fat since these cyclic constituents are specific to irradiation treatment. They form at levels below 0.5 kGy and can be detected at low concentrations by gas chromatography combined with mass spectrometry (GC-MS).

4.3. Combined treatments

Some of the stabilization processes described here result in a partial degradation of components in the treated products, which may be inevitable or uncontrolled. In the case of shelf-stable products, it is important to inactivate not only the microbial flora but also enzymes. It is impossible to denature certain endogenous enzymes in the product to complete stabilization (e.g. milk plasmin) without affecting other components due to treatment conditions (serum proteins, Maillard reaction, loss of lysine).

In order to ensure complete safety of food, especially heat-sensitive products (egg whites), and improve its nutritional and sensory quality, food industries have, in recent years, sought to develop effective processes that are less denaturing than heat treatments. A variety of low-heat physical preservation or processing methods (microfiltration, high hydrostatic pressure, pulsed electric fields, pulsed light, etc.) have been developed, often as part of national or European contracts. In conjunction with this, new heat treatment techniques (ohmic heating, steam injection UHT, etc.) were also proposed. Finally, another prospective method involves coupling two techniques to achieve minimal processing conditions. The effects of these different emerging technologies on the elimination or inactivation of microorganisms and enzymes, and on the physicochemical, biochemical, nutritional and toxicological properties of major food components are still not well known, despite a large number of publications. The implementation of the European Directive on "Novel Foods" since 1997 requires a thorough knowledge of the changes in foods that have been subjected to any innovative technology.

One example worth mentioning here is milk stabilized by cross-flow microfiltration, alone or combined with heat treatments. Microfiltration alone can be used to obtain drinking milk with its original taste intact and a use-by date range of 21 days. When combined with moderate heat treatment, and depending on its intensity, this range extends from 35 days (combined with a treatment of 20 s at 72°C) to 6 months (treatment of 6 s at 96°C). This type of treatment is hardly used in France at present, but has seen remarkable growth in North America.

PART 3

Food Physicochemical Stabilization

5

Stability of Complex Foods and Dispersed Systems

5.1. Complex foods: overview of dispersed systems

5.1.1. *Definitions*

5.1.1.1. *Emulsions*

An emulsion is a mixture of two or more liquids that are normally immiscible; it is often necessary to blend two such liquids so that the mixture can be used without any phase separation. One of the most common techniques is to emulsify one phase in the other by combining mechanical agitation in the presence of an emulsifier, which stabilizes the interface. The result is an emulsion, which can usually be described as a dispersion of droplets of one phase in the other. A distinction is therefore made between the dispersed phase and the continuous (dispersing) phase. If the continuous phase is an oil phase, this is referred to as a water-in-oil emulsion, and if the continuous phase is a polar liquid (usually water or an aqueous solution), this is referred to as an oil-in-water emulsion. In general, the volume fraction of the continuous phase is higher. When the continuous phase is lower in volume (less than 30%), this is known as a concentrated emulsion. The size of the droplets forming the dispersed phase of an emulsion is in the micrometer range (0.1 to 100 µm) and appears opaque white in most

---

Chapter written by Romain JEANTET and Juliane FLOURY.

emulsions. An emulsion can be fluid, viscous or gelled resulting in a whole range of textures, which explains the interest of the agri-food and cosmetic industries in such systems [BRO 99].

5.1.1.2. *Whipped emulsions and foams*

Edible foams are usually dispersions of gas bubbles in a liquid or semi-solid continuous phase, stabilized by the addition of amphiphilic molecules called surfactants (Figure 5.1). Whipped emulsions are more complex products since a gas phase and a liquid or semi-solid phase are dispersed in a liquid continuous phase. This dispersed system therefore consists of three phases, liquid–liquid–gas or liquid–solid–gas, which can generate a wide variety of textures, generally highly valued for their "lightness".

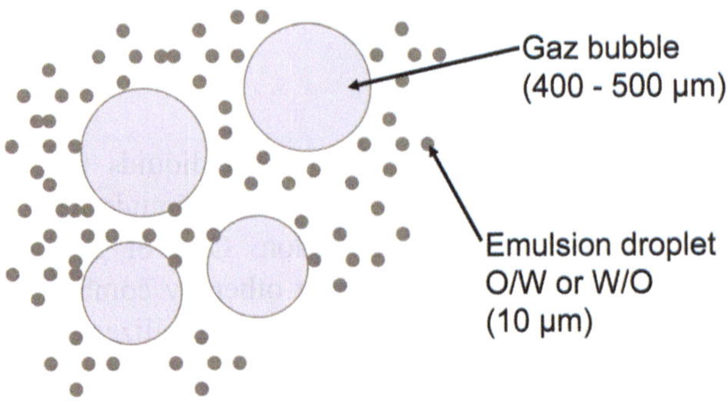

Figure 5.1. *Whipped emulsion.*
O/W: oil-in-water; W/O: water-in-oil

There are a wide variety of food foams such as meringues, cakes, marshmallow, whipped cream, ice cream, soufflés or bread. In many cases, the gas is air (or possibly carbon dioxide) and the continuous phase is a suspension or an aqueous emulsion containing protein. Some food foams are more complex colloidal systems. Ice cream, for example, contains a suspension of dispersed ice crystals, air bubbles and fat globules (mostly solid and grouped in small clusters) in a concentrated solution of sugars and proteins, all of which are trapped in a polysaccharide matrix [CHE 85].

5.1.2. *Emulsion stability*

An emulsion is a thermodynamically unstable system involving two immiscible phases that tend to minimize their contact area. Emulsion stability relies on slowing or preventing the physical mechanisms that naturally lead to the separation of immiscible phases.

There are three types of mechanisms of emulsion instability (Figure 5.2):

– sedimentation or creaming;

– flocculation or droplet aggregation;

– coalescence or fusion of oil droplets.

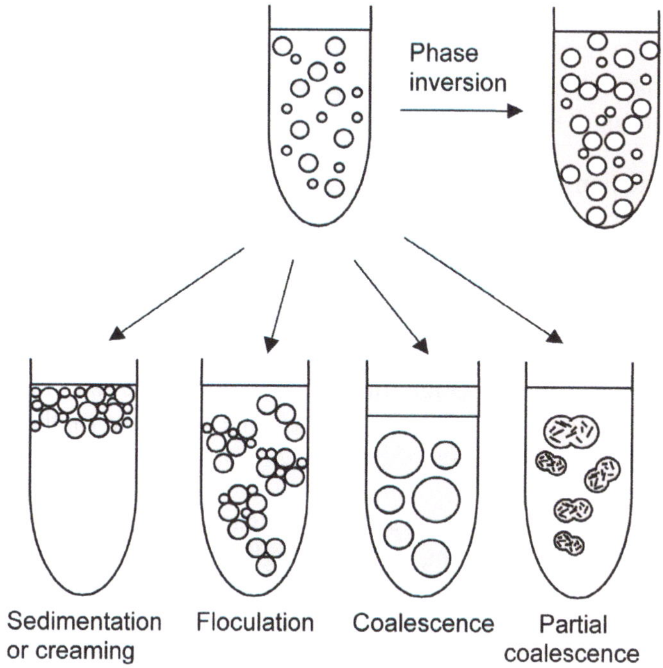

Figure 5.2. *Mechanisms of emulsion instability*

5.1.2.1. *Sedimentation or creaming*

Creaming is due to the movement of oil droplets in an oil-in-water emulsion as a result of gravity or a centrifugal force, and leads to the formation of a concentrated layer above the emulsion without a change in the droplet size distribution. Initially a vertical gradient in droplet concentration occurs, but subsequently a separate layer may appear between the top layer of cream and the layer below, depleted of droplets. In general, creaming is reversible since the initial distribution of droplets in the dispersing phase can be restored by gentle agitation.

The velocity of an isolated droplet in a Newtonian liquid is given by Stokes' law [3.1]:

$$\mathrm{v} = \frac{D^2}{18\eta_c}(\rho_d - \rho_c)\,g$$

where v is the settling velocity (m s^{-1}), D is the droplet diameter (m), $[\rho_d - \rho_c]$ is the density difference between the two phases (kg m^{-3}), η_c is the dynamic viscosity of the continuous phase (Pa s) and g is the gravitational acceleration (m s^{-2}).

This equation is applicable only under limited conditions. A number of factors affect the creaming rate of an emulsion: the volume fraction of the dispersed phase, the droplet size distribution, the hydrodynamic and interfacial properties of droplets as well as the rheological behavior of the continuous phase. In the absence of flocculation, Stokes' law shows that there are three ways to prevent creaming in a diluted emulsion: reduce the average droplet size, decrease the mass density difference between the phases or increase the viscosity of the dispersing medium.

In general, food emulsions are stable against creaming if the average droplet size is less than 1 μm, although slight creaming is inevitable due to the existence of droplets larger than this size (Gaussian distribution of sizes). Low creaming rates are permissible if there is no change to the texture or appearance of the emulsion [DIC 92].

5.1.2.2. *Flocculation or aggregation*

When emulsions are aggregated, the droplets formed do not remain independent of each other but tend to group together to form clusters. The droplets remain very close to each other for a long period of time without a rupture of the film separating them. This phenomenon is known as flocculation.

The aggregation of droplets results in an increase in the apparent size of particles, and therefore, in most cases, leads to decreased stability and creaming of emulsions (increase in diameter D in Stokes' law). A distinction is often made between the different types of aggregation depending on their degree of reversibility by using terms such as aggregation, flocculation or agglomeration. However, since these terms are difficult to distinguish, the general term of aggregation will henceforth be used.

The DLVO theory (Derjaguin, Landau, Verwey, Overbeek) describes the stability of colloidal suspensions with regards to agglomeration, coagulation or flocculation. It establishes the balance of forces affecting particles. There are two types:

– van der Waals forces: they are attractive and mainly result from dipolar interactions at molecular level. The potential energy of attraction depends on the nature of the material and the dispersion medium, particle size and the inter-particle distance.

– electrostatic forces: they are repulsive and result from the surface potential of charged particles and the repulsion between the electric double layer at the surface of two approaching particles.

The free energy of interaction (G_T) is the sum of van der Waals attractions (G_W) and electrostatic repulsions (G_E). The droplets of an emulsion aggregate if the force of attraction is greater than the force of repulsion, that is if the free energy of interaction is negative. This free energy of interaction, which depends on the distance between droplets, often has two minima (Figure 5.3).

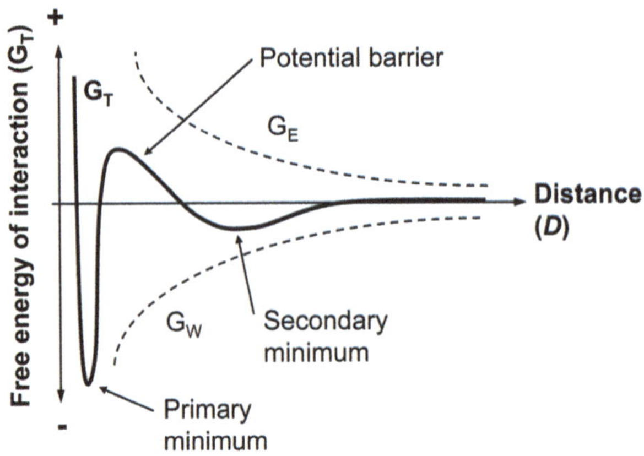

Figure 5.3. *Variation in potential energies as a function of inter-particle distance*

At close distances, the van der Waals attraction between droplets is very strong, and a first minimum, called the primary minimum, is observed. The van der Waals attraction then rapidly decreases with increasing inter-droplet distance whereas the electrostatic repulsion can remain relatively strong resulting in maximum free energy of interaction. If the inter-droplet distance increases further, the electrostatic repulsion and, to a lesser extent, the van der Waals attraction decrease, often leading to a secondary minimum of free energy of interaction.

The presence of surfactants adsorbed at the droplet interface often affects the relationship between the free energy of interaction and the inter-droplet distance, and therefore renders the interaction between attractive and repulsive forces more complex. Thus, polymeric emulsifiers adsorbed at the droplet interface often result in a significant repulsion between droplets by forming a thick layer of polymer chains. If the adsorbed polymers are polyelectrolytes, the aggregation of droplets takes place below a certain threshold of dissociation of acid-base groups, which depends on pH and ionic strength. Strong aggregation is also possible when the polymeric emulsifiers have reactive sites capable of forming covalent bonds with adsorbed polymers on neighboring droplets. Attractive electrostatic interactions between oppositely charged groups of adsorbed polymers can also cause the aggregation of an emulsion.

Other possible types of aggregation are bridging flocculation and depletion flocculation. Bridging flocculation is caused by polymers of high molecular weight that adsorb to the surface of more than one droplet and link them together. It occurs when the concentration of polymers is too low to saturate the droplet surface. In this case, polymers and surfactant molecules are simultaneously adsorbed on several droplets, thereby allowing bridging and consequently flocculation. Bridging flocculation generally occurs in emulsions of low droplet packing density.

Depletion flocculation occurs when high molecular weight polymers are present in the continuous phase of an emulsion of high droplet packing density. When two lipid droplets are close to each other, these polymers, initially distributed throughout the available volume of the continuous phase, are excluded from the space between the droplets; this area of continuous phase becomes too small to allow the free motion of polymers leaving the narrow region surrounding the droplet surface: the result is depletion. As a result, the polymer concentration increases more in the continuous phase than in the proximity of the droplet interface, leading to an increase in osmotic pressure in the continuous phase. The aggregation of droplets lowers this pressure by reducing the perimeter of the depletion region. Depletion flocculation can occur with non-adsorbed polymers, like polysaccharides, or with small surfactant particles that form micelles.

5.1.2.3. *Coalescence*

Coalescence is the fusion of two or more emulsion droplets to form a single larger droplet. Two droplets aggregated by flocculation are still separated by a liquid layer that needs to be ruptured for coalescence to occur. Coalescence thus depends on the stability of the thin liquid layer of continuous phase separating the two droplets. Generally, coalescence occurs when the droplets are very close to each other for a prolonged period. Rupture of the interstitial film occurs when it drops below a critical thickness.

Deformation leading to the rupture of the interface is slowed down by the viscoelasticity of the surface, in which the physicochemical properties of the emulsifier and its concentration play a major role. Coalescence can be caused by a delay in the adsorption or desorption of the emulsifying agent during changes to the interfacial area or a local depletion of the emulsifier, which both result in interface destabilization, therefore leading to a rupture of the droplets (Gibbs–Marangoni effect). The theory in section 5.2.1.1 predicts that the film rupture is favored by a large radius of curvature of the film (i.e. a large droplet diameter), weak repulsion between droplets and low interfacial tension.

Based on observations and theoretical knowledge, physical chemists and emulsion specialists have identified a number of general factors that seem to correlate with emulsion stability. Table 5.1 summarizes and highlights the relative importance of 12 key parameters related to creaming, flocculation and coalescence. A separate section is attributed to rheology since a close link has been found between the stability and rheology of emulsions.

In the case of partial coalescence, two or more partially crystallized oil droplets come into contact and form an irregularly shaped aggregate (Figure 5.2). The aggregate maintains the shape of the droplets from which it was formed since the fat crystal network imposes mechanical constraints that prevent the droplets from merging completely. Partial coalescence only occurs in emulsions containing partially crystallized droplets. If the droplets were completely liquid, normal coalescence would occur, and if they were completely solid, flocculation rather than partial coalescence would occur since the droplets would not be able to merge. In addition, partial coalescence is affected by most of the factors that influence normal coalescence. It is induced by agitation but is faster than normal coalescence (liquid fat) due to the protrusion of fat crystals that pierce the film of the droplets coming into contact with them. Partial coalescence is essential in the production of whipped cream and ice cream from milk. In these systems, which are aerated emulsions, air bubbles are stabilized with regard to coalescence by forming a thick composite layer of partially coalesced fat, proteins and emulsifiers at the air bubble surface.

Factor	*Creaming*	*Flocculation*	*Coalescence*	*Rheology*
Droplet size	+++	++	+	+
Droplet size distribution	+++	++	-	++
Droplet volume fraction	+++	+++	+++	+++
Mass density difference between the phases	+++	-	-	-
Rheology of the continuous phase	+++	+++	++	+++
Rheology of the dispersed phase	-	-	-	+
Rheology of the adsorbed layer	-	-	+++	++
Thickness of adsorbed layer	+	++	+++	++
Electrostatic interactions	+	+++	++	+
Steric interactions (polymeric)	-	+++	++	++
Fat crystallization	-	-	+++	+++
Liquid crystalline phases	+	+	++	++

Table 5.1. *Twelve key physical parameters affecting the stability and rheology of emulsions: no effect; + little effect; ++ medium effect; +++ significant effect [DIC 92]*

5.1.3. *Foam stability*

In foams, the thin liquid layer separating two bubbles is known as the lamella, and the junction between the lamellae of adjacent bubbles is known as the Plateau border (Figure 5.4).

In foams, the volume fraction occupied by the dispersed phase (gas) varies much more than with emulsions. Foams are thermodynamically unstable because they have a very large interfacial area and the mass density difference between the continuous phase and the dispersed phase is very large. If the volume of gas represents more than 74% of the volume of the foam, it is not possible that the latter is composed of spherical bubbles alongside each other; they are hexagonal.

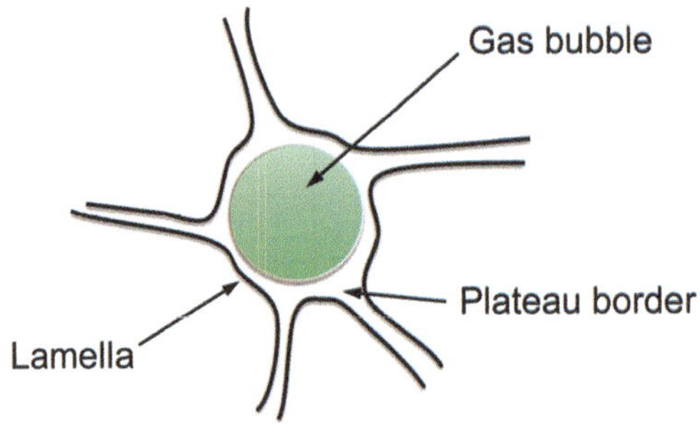

Figure 5.4. *Diagram of the structure of a foam*

There are three main destabilization mechanisms with regard to foams:

– Ostwald ripening or disproportionation;

– drainage or flow of liquid by gravity (across the Plateau border) or by capillary action (lamellae to the Plateau border);

– desorption of the foaming agent under the effect of excessive shear or heat.

5.1.3.1. *Ostwald ripening*

Ostwald ripening corresponds to the growth of large bubbles at the expense of smaller ones by the diffusion of gas through the continuous phase (lamellae). This is due to a higher gas pressure in the small bubbles, according to Laplace's equation:

$$P_i - P_e = \frac{2\sigma}{R} \quad [5.1]$$

where P_i and P_e are the pressure inside and outside the bubble, respectively, σ is the interfacial tension (N m^{-1}) and R is the radius of the bubble.

This phenomenon is known as Ostwald ripening. It does not occur in emulsions due to the low solubility of triglycerides in the continuous aqueous phase. It can, however, occur in emulsions that are very rich in proteins, which can act as carriers of fatty acid molecules, or when the continuous phase is an aqueous-alcoholic solution, the alcohol improves the solubility of the triglycerides.

5.1.3.2. *Drainage*

Due to the effect of gravity, the solution surrounding the air bubbles drains, leading to a thinning of the films. In low-density foams, bubbles tend to press tightly against each other; they deform and become hexagonal, producing pressure gradients between the lamellae and the Plateau border, which increases the flow of liquid. Low interfacial tensions and large bubble diameters decrease internal pressure and flow equation [5.1]. Flow is also reduced when the liquid phase is viscous (this can be obtained by adding sugar for example) and when the viscoelasticity of the interface is high. In this respect, protein foams are more stable with regard to drainage than foams made of low molecular weight surfactants.

5.1.3.3. *Desorption of the foaming agent*

A rupture of the liquid films and the coalescence of bubbles by the desorption of the foaming agent, under the action of excessive shear and/or heat, ultimately results in a collapse of the foam.

The gas contained in the bubbles can also be released under the effect of heat, resulting in the destabilization of the foam.

In the case of aerated emulsions, the disintegration of the foam is often linked to the instability of the liquid–liquid emulsion itself. The three main factors that contribute to stabilizing aerated emulsions are:

– low interfacial tension;

– high viscosity of the liquid phase;

– elastic and resistant adsorbed protein films.

5.2. Production of emulsions

5.2.1. *Fractionation/coalescence*

5.2.1.1. *Energetic and thermodynamic aspects*

Emulsions are not formed spontaneously. Producing a stable emulsion requires a large amount of energy. The free energy needed to increase the oil-water interface by an area ΔA is:

$$\Delta G = \sigma \, \Delta A \qquad [5.2]$$

where σ is the interfacial tension ($N\,m^{-1}$), defined by:

$$\sigma = \left(\frac{\partial G}{\partial A}\right)_{P,\,T,\,n} \qquad [5.3]$$

with G the Gibbs free energy of the system (Nm), A the surface area (m^2), *P* pressure (Pa), T temperature (K) and n the total quantity of material in the system (mol).

In practice, the actual amount of energy required to obtain a stable emulsion is at least a thousand times higher. This apparent excess energy is needed to form small droplets from larger ones. The rupture of a spherical droplet of radius *R* requires the application of an external pressure gradient:

$$\frac{dP}{dx} \cong \frac{\Delta P}{R} = \frac{2\sigma}{R^2} \qquad [5.4]$$

where ΔP is the Laplace pressure defined by equation [5.1]. Such a local pressure gradient is generally obtained using vigorous agitation.

It is difficult to produce emulsions without surfactants. Surfactants act by lowering the interfacial tension, which leads to a decrease in Laplace pressure for the same droplet size. [5.4] shows that the decrease in interfacial tension σ by the adsorption of the surfactant on the droplet surface decreases the level of agitation to obtain the same droplet size. Surfactants are also necessary for the formation of the interfacial film that resists the destabilization mechanisms described above (section 5.1.2). When there is sufficient surfactant to completely cover the interface, the energy input by the device used to produce emulsion is the limiting factor for the size of emulsion droplets.

Apart from the very small amount of energy required to create the interface, the excess mechanical energy is dissipated as heat. From a thermodynamic perspective, the yield of the emulsification process is very low.

5.2.1.2. *Droplet breakup*

In general, droplet breakup can be seen as a process involving two opposing forces: an external force induced by flow, which tends to deform the droplet, and an internal restoring force, which tends to maintain the initial droplet shape (Laplace). Rupture occurs when the deforming force becomes greater.

These pressure gradient phenomena exerted on droplets may occur homogeneously in the laminar shear field (where the viscosity of the medium is high), or more locally in the boundary layer at the wall in turbulent flow (in the case of liquids with low viscosity).

Purely viscous shear stresses dominate in laminar flow and rupture can be modeled by the capillary number, the ratio between the viscous deforming force and the internal restoring force. Droplet fractionation occurs if the capillary number is greater than a certain value known as the critical capillary number:

$$\mathrm{Ca} = \frac{\dot{\gamma}\ \eta_c\ D}{2\sigma} \qquad [5.5]$$

with $\dot{\gamma}$ being the shear rate (s^{-1}), η_c the viscosity of the continuous phase (Pa s), D the droplet diameter (m) and σ the interfacial tension ($N\ m^{-1}$).

In the case of turbulent flow, inertia combined with local velocity fluctuations cause the stresses. Rupture in this case can be modeled in the same way as before using the Weber number (see Chapter 1, section 1.2.4.1):

$$W_e = \frac{\rho_c \, \Delta v^2 \, D}{\sigma} \qquad [5.6]$$

where Δv is the velocity of the liquid close to the droplet and $\rho_c \Delta v^2$ is the pressure difference created by the turbulent eddies. The flow conditions are usually difficult to accurately define because of the complexity of the droplet shapes used.

These dimensionless numbers describe three key parameters that determine the size limit of dispersed elements:

– the viscosity of the continuous phase;

– the agitation rate on which the local shear rate depends;

– the interfacial tension.

5.2.1.3. *Role of surfactants*

Lowering interfacial tension

The lowering of the interfacial tension σ between the dispersed phase and the continuous phase is one function of a surfactant during emulsification. The interfacial tension between most oils and water varies between 25 to 30 mN m^{-1}. The addition of surfactants can bring this value down to 10 mN m^{-1} (in the case of proteins). Effective emulsifiers can reach 1 mN m^{-1} (in the case of phospholipids), and certain surfactant/cosurfactant mixtures used in high concentration yield interfacial tensions of a few µN m^{-1}.

However, all these values are given at equilibrium. During emulsification, the interfacial tension on the droplet surface can be much higher than the equilibrium value. In addition, the adsorption time can be relatively long compared to the rupture time of the droplets and the collision time. Therefore, lowering the interfacial tension is not necessarily the primary role of surfactants during emulsification.

Effect of deformation stress

Droplet interface is subjected to shear stress due to fluid flow along the interface. This creates variations in the surfactant concentration at the droplet interface and consequently interfacial tension gradients, which delay the tangential motion of the interface, and thus the fluid flow along this interface. In the absence of surfactant, fluid flow is continuous on both sides of the interface. In the presence of surfactant, fluid flow along the interface creates a surfactant concentration gradient, which in turn induces a migration of the surfactants at the interface opposite to the fluid flow in order to reduce the concentration gradient. If high enough, these gradients (surfactant concentration and interfacial tension) can immobilize the surface and the associated continuous phase. In addition, surfactants limit the flow of liquid at the centre of the droplet. Surfactants can also create a viscoelastic interfacial layer, thereby improving the mechanical resistance of the interface.

It is clear that these effects impact the rupture of droplets as well as their size. During emulsification, shear conditions (shear rate $\dot{\gamma}$) vary considerably, and it is particularly difficult to make predictions. However, the presence of a surfactant aids in the formation of smaller droplets.

Prevention of coalescence

Most of the stages of emulsification, that is the formation of the interface, the adsorption of the surfactant, its spreading at the interface and the collision of droplets, occur within a few milliseconds or less. In order to obtain a stable emulsion, the time interval between the creation and collision of small droplets should be long compared to the time necessary for the adsorption of the surfactant and the formation of a stabilizing layer at the oil/water interface. In the case of low molecular weight surfactants, stabilization with regard to coalescence during emulsification is attributed to the Gibbs-Marangoni effect.

The adsorption of polymers at the oil/water interface, controlled by diffusion, is slower than that of low molecular weight surfactants. Thus, using the latter in combination with proteins can facilitate the rapid lowering of interfacial tension and therefore the formation of a food emulsion. Changes in interfacial tension, by varying the composition and chemical nature of the surfactants, do not significantly reduce droplet size compared to the effects of variations in energy input (the latter can vary by several orders of magnitude).

Some authors believe that coalescence occurs during the emulsification process (Figure 5.5), but that it mainly involves newly-formed droplets that are not yet stabilized by an interfacial layer at equilibrium [WAL 83, KAR 95, WAL 98]. At present, the role of coalescence in the emulsification process still remains unknown. It could partially explain why there are different particle sizes depending on the type of surfactant used.

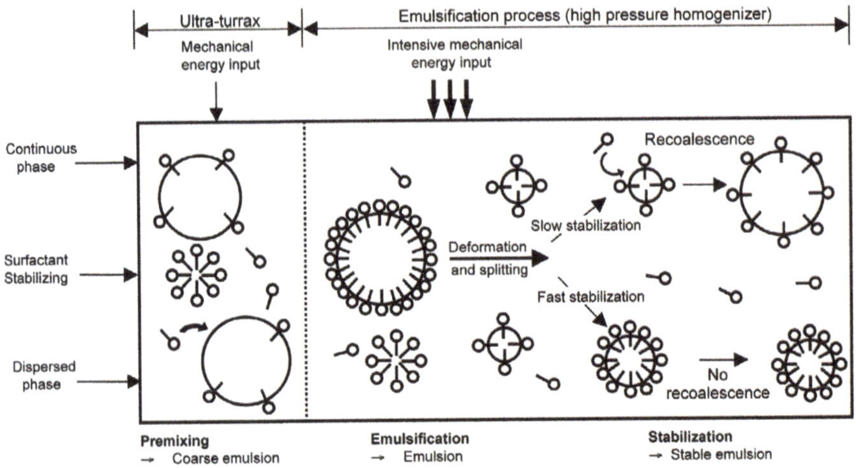

Figure 5.5. *Diagram of emulsification process (according to [KAR 95])*

5.2.2. *Practical aspects of emulsification*

5.2.2.1. *Emulsification process*

Regardless of the emulsification process chosen, the rupture of dispersion elements requires sufficient shear to obtain the most effective dispersion of the lipid phase, or in other words, to obtain the smallest possible oil droplets. Many devices have been designed to produce emulsions (Figure 5.6). Of all these techniques, some are used solely in the laboratory (shakers, vibrators, magnetostriction, aerosols, etc.). The most commonly used devices in industrial applications are rotor-stators, colloid mills, high-pressure homogenizers ($P < 100$ MPa) and ultrasound systems. Most of the time a combination of two techniques is used, for example agitation together with high-pressure homogenization.

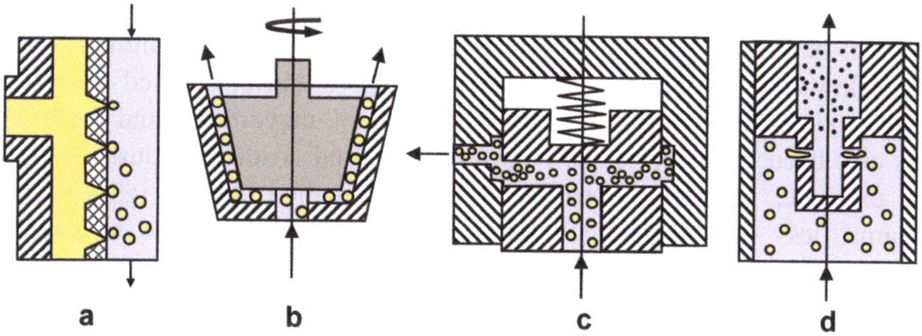

Figure 5.6. *Continuous emulsification equipment (according to [SCH 97]): a) membrane; b) colloid mill; c) high-pressure homogenizer; d) jet dispersion system*

Mixers

Emulsions can be produced in a tank with a simple agitator. A high-speed rotating blade in a stator is a commonly used device. However, even if batch processing is preferred, the operation is much more effective in reducing droplet size if shear is applied in the narrow gap between rotor and stator. This is why, in high shear mixers like the "Ultra Turrax", the rotor ring rotates between two slotted rings. The product is drawn into the bottom of the emulsification head and expelled through the slots of the outer ring of the stator.

High-pressure homogenizers

High-pressure homogenization involves forcing a viscous emulsion, using a high-pressure pump, through a butterfly valve with an adjustable opening to reduce the size of the fat globules (Figure 5.7). This dispersion process, in which the energy density is determined solely by pressure, is particularly suitable for the production of very fine emulsions of any viscosity. With energy densities equal to other conventional emulsification systems (rotor-stator, colloid mills), high-pressure systems are more effective than the latter because the mean residence time is shorter, and therefore the power density is higher. The cavitation and elongation forces in high-pressure systems are favorable for the production of smaller droplets using the same energy density.

In 1890, Gaulin started using this technique for the homogenization of milk. It has the advantage of being able to treat products continuously at high flow rates. This homogenization process is currently used for many food products such as ice cream, yoghurt, cream, mayonnaise and sauces. It has also been extended to the pharmaceutical and cosmetic industries with, for example, parenteral nutrition fluids or so-called "lightweight" beauty creams (less oily appearance) allowing better hydration due to smaller droplets.

In order to develop new products, improve the quality of existing products and reduce production costs, new high-pressure valve homogenizers have been proposed. Due to highly resistant ceramic valve systems and very powerful pressure multipliers, it is possible to achieve much higher homogenization pressures (up to 400 MPa) than those normally used in the food industry (between 30 and 100 MPa for conventional homogenizers).

Other homogenization methods

In colloid mills, designed for continuous processes, the product passes through the narrow gap between rotor and stator where it is subjected to high shear (Figure 5.7). In addition to the depth and direction of the rotor and stator serrations, which vary from one device to another, the process parameters are the product flow rate, rotor speed and the size of the gap. Colloid mills are more suitable for highly viscous liquids than high-pressure homogenizers, reserved for relatively low viscosity liquids. Like the homogenizer, it has the advantage of being able to treat products continuously at high flow rates, but it can cause considerable heating of the fluid in the device. The emulsions obtained are relatively fine, with an average droplet diameter of around 2 to 4 μm.

New methods have recently emerged, which include jet and membrane emulsification systems (Figure 5.7). The jet system is based on the collision of two pre-emulsified streams in a chamber, at the end of which an emulsion is obtained. With this type of homogenizer, known as a microfluidizer, increasing the speed of the two jets increases the homogenization pressure, which can reach 100–150 MPa with some devices. The mechanisms involved are turbulence, shear elongation forces and cavitation. Membrane emulsification involves forcing the dispersed phase through the pores of a microfiltration or ultrafiltration membrane into the continuous phase. The droplets thus formed are detached from the membrane surface by flow of the continuous phase. The mechanisms of droplet formation are very different to conventional processes. The size and size distribution of droplets are primarily determined by the size and size distribution of the membrane pores. This process has the major disadvantage of a sudden increase in pressure drop when flow increases and the emulsion becomes concentrated.

5.2.2.2. *Effect of temperature increase*

During the sudden pressure drop in the valve of a high-pressure homogenizer, almost all of the kinetic energy is dissipated by viscous stress in the form of heat. Thus, high-pressure homogenizers produce a maximum heat of about 20°C/100 MPa.

The temperature increase associated with the emulsification method used is an important parameter to be taken into account in the production of an emulsified product. A considerable increase in product temperature can affect the properties of the emulsion ingredients (heat denaturation of macromolecules) as well as the size of oil droplets.

The effect of temperature in processes such as high-pressure homogenization is still poorly understood given the very short residence time of the fluid in the homogenizing valve. The consequences of this almost instantaneous temperature increase during dynamic high-pressure treatment are likely to be very different from those of a conventional heat treatment corresponding to a very specific time–temperature relationship.

Oil-in-water emulsions are generally insensitive to small changes in temperature. However, the effectiveness of homogenization is improved by heating, which results in a partial melting of triglycerides in the oil phase. Several authors have shown that in the case of butter fat, a temperature increase of 10°C between 40 and 70°C lowered the average fat globule diameter by 6–8% [WAL 75] or 10–15% [SWE 83]. This effect decreases or disappears above 80°C, beyond which the mean droplet diameter increases slightly due to protein denaturation [ROB 93].

It is a well-established fact that a temperature increase reduces interfacial tension between phases, as well as their viscosity, according to an Arrhenius-type law. This dual action facilitates the division of the dispersed phase and the decrease in the average fat globule diameter. However, the higher the temperature is, the greater the kinetic energy of the droplets and the higher the inter-particle collision frequency. In addition, an excessive temperature increase may degrade certain constituents, such as proteins, and modify the interfacial adsorption of emulsifiers. As a result, if excessive heating occurs during the homogenization process, particles tend to coagulate, which can lead to a modification of the emulsion properties.

5.2.2.3. *Dynamic high-pressure treatment*

The development of the homogenization process at very high pressures has resulted in a greater interest in emulsion stability. However, these high pressures not only affect the characteristics of the oil droplets in an emulsion, but also the properties of other food components such as polymers or colloids:

– protein denaturation and modification of their functional properties (modification of gel-forming ability, thickening and emulsifying properties);

– reduction and homogenization of the size of protein aggregates formed during heat treatment. For example, heat-denatured and micro-particulated whey proteins are sometimes added to milk intended for cheese production to improve the water retention capacity of cheeses. Spherical particles with diameters of a few μm can also be used to replace fat globules in low-fat products.

– structural modification of polysaccharides and in some cases a reduction in polymer size.

5.3. Stability of dispersed systems

An emulsion such as mayonnaise that was frozen, defrosted and then heated in a microwave will lose its structure and collapse: the fat separates from the aqueous phase. This example shows that the different stresses applied to food by manufacturers, distributors or consumers can impact its physical structure. In many cases, the food industry is able to overcome these challenges by using polysaccharides, proteins or low molecular weight surfactants, which create, modify and stabilize the microstructure of food products. These molecules function during processing (thermal, mechanical) and/or during storage.

5.3.1. *Polysaccharides and proteins*

Thickening and gelling agents are generally polysaccharides or proteins that easily dissolve or disperse in water resulting in an increase in viscosity and sometimes the formation of a gel under the effect of physical (temperature) and/or chemical (presence of ions, change in pH) and/or biological factors (enzyme action). Due to their thickening and/or gelling ability, these polymers can be used to stabilize suspensions and dispersions (emulsions, foams). They are also used for their water retention capacity, binding ability, texturing role, etc.

5.3.1.1. *Origin, properties and applications*

Polysaccharides

Of the polysaccharides used in the food industry, starches are undoubtedly the most common. The use of native starches poses a number of disadvantages for changes that occur either during processing or in terms of the sensory characteristics of the final product. For example, some starches such as those in potatoes, cassavas or "waxy" corn develop a "sticky" texture in the final product, which is unappealing to the consumer. Similarly, dispersions of high-amylose starch, such as native corn or wheat starch, form opaque and rigid structures upon cooling due to retrogradation. Some or all of these disadvantages can be overcome by:

– strengthening the cohesion of starch granules by chemical cross-linking to ensure good stability at low pH, at high temperatures and during shearing. Cross-linked starch in food products includes distarch adipate and distarch phosphate;

– creating steric hindrance or electrostatic repulsion by etherification or esterification to limit the risk of retrogradation.

Apart from starch, other polysaccharides are also widely used in the food industry (Table 5.2).

<table>
<tr><th>Type</th><th>Origin and structure</th><th>Physicochemical and functional properties</th><th>Application</th></tr>
<tr><td>Agar</td><td rowspan="2">Red algae of the Rhodophyceae family.
Linear polymers whose repeating unit is composed of two galactose residues that are partly substituted.</td><td>Soluble at very high temperatures (boiling).
Gelation with firm, brittle, thermally reversible gel (similar to carrageenan).</td><td>Preparation of traditional Asian desserts, confectionery.</td></tr>
<tr><td>Carrageenans (E407)</td><td>Fraction rich in λ-carrageenan: cold water soluble thickener.
Fraction rich in ι-carrageenan: gelling agent with elastic and cohesive gel.
Fraction rich in κ-carrageenan: gelling agent with firm and brittle gel (with some syneresis).
Synergy of κ fractions with carob.
Strong interaction with proteins.</td><td>Dairy products (stabilization of chocolate drinks and cream (UHT – aerosols – whipped), gelled milk, dairy desserts, mousses.
Powdered products (cake mix, cream desserts, custard, pastries, jellies and topping gels).
Ice cream, meat and fish, preserves.</td></tr>
<tr><td>Alginates (E401)</td><td>Brown algae of the Phaeophyceae class.
Linear polymers composed of two 1,4-linked monomers: β-D-mannuronic acid and α-L-guluronic acid.</td><td>Cold water soluble thickener in water low in calcium.
Gelling agent in the presence of calcium and acid.
Non-thermally reversible gel.</td><td>Dairy products (cream, whipped cream, dairy desserts, processed cheese, ice cream).
Powdered products (Pastry cream, bechamel).
Restructured products (meat, fruit, vegetables, fish).</td></tr>
<tr><td>Carob (E410)</td><td>Grain legumes, e.g. from the Mediterranean carob tree.
Linear chain of β-(1,4)-linked D-mannose units with single-unit branches of α-(1,6) linked D-galactose.</td><td>Thickener soluble after heating.
Gelling agent in combination with xanthan gum or κ-carrageenan.</td><td>Dairy gels (with carrageenan).
Ice cream (with carrageenan and alginates).
Canned pet food.
Fruit preparations (with pectin).
Bread and pastries.</td></tr>
</table>

Guar (E412)		Cold water soluble thickener. Good freeze-thaw resistance.	Acidic and neutral dairy products, sauces, ice-cream, sorbets, bread, pastries, powdered products.
Pectins (E440)	Complex mixture of polysaccharides extracted from the cell wall of plants. Major constituent: (1,4)-linked galacturonic acid, rhamnose groups and homogalacturonic regions	HM pectin: gelling agent in acidic and sugar medium, non-thermally reversible gel. LM pectin: gelling agent in the presence of calcium, thermally reversible gel.	Dairy products (stabilization of acidic dairy beverages (LM), yoghurts (HM), fruit preparations for yoghurts, fruit preparations for dairy dessert), standard and low-sugar jams (LM), fruit preparations for yoghurt, preparations for pastries/biscuits (toppings, fillings), fruit drinks, confectionery (fruit paste)).
Gum arabic (acacia) (E414)	Acacia Senegal. Highly branched polymer: main chain contains β-(1,3)-linked D-galactose residues.	Forms a protective colorless and odorless film that retains aroma. Rich in fiber, water-soluble and low in calories.	Soft drinks (concentrated aromatic emulsions known as "soft drink concentrates), confectionery (gumballs, lozenges and coated products). Dietetic preparations (diet products). Extruded products.
Xanthan gum (E415)	Synthesized by microorganisms. Aerobic fermentation of *Xanthomonas campestris*. Linear chain of β (1,4)-linked D-glucose with a side branch at every second glucose unit.	Cold water soluble thickener. Can easily form suspensions. Remarkable stability in acid medium. Good freeze-thaw resistance. High melt viscosity. Gelling agent even in the presence of carob flour.	Salad dressings (stability in acid medium). Sauces and dairy products (melt viscosity). Powdered products (rapid dissolution).

Gellan gum	Synthesized by microorganisms. Aerobic fermentation of *Pseudomonas elodea.* Native form: linear polymer. Repeating unit is a tetrasaccharide of two D-glucose (every 2nd unit) with glucuronic acid and L-rhamnose alternating in between.	In its native form, it forms weak, elastic and thermally reversible gels. Can form gels with a wide range of properties by controlling the degree of acylation. Good flavor and stability for large range of pH values.	Dairy products, spreads, salad dressings and sauces, which can withstand freezing, UHT treatment or treatment in a scraped surface heat exchanger.
Cellulose derivatives (E460 - 466)	Cotton cellulose or wood products Cellulose: linear polysaccharide insoluble in water, composed solely of β (1,4)-linked D-glucose. Obtained by chemically modifying cellulose by etherification in order to make it water-soluble.	Cellulose derivatives: thickening and binding agents in foods (Carboxymethyl cellulose E466, Methyl cellulose E461). Other cellulose ethers: Hydroxypropyl methyl cellulose E463 Ethyl methyl cellulose E465 Hydroxypropyl methyl cellulose E464.	Used in sauces as thickening and binding agents or as stabilizers in products like ice cream.

Table 5.2. *Main texturizing agents (other than starch) and their food applications*

[DER 09] provide additional information on the chemical structure, functional properties and food applications of polysaccharide thickening and gelling agents. Further information is available from suppliers of ingredients.

Proteins

One of the most widely used proteins is gelatin due to its extensive range of functionalities (thickening, gelling, emulsifying and foaming properties). It is obtained by the partial hydrolysis of collagen in the skin, connective tissue and bones of animals. With the exception of tryptophan, gelatin contains all the amino acids, with a large amount of glycine, proline and hydroxyproline.

Globular proteins are widespread in nature. They are found in large quantities in egg white, whey and plant products like soybean. They have a relatively compact structure and are soluble across a wide pH and ionic strength range. However, they are sensitive to temperature and exposure to hydrophobic interfaces (oil, air), conditions that are widely used in the production of thermotropic gels and dispersions (foams and emulsions).

Caseins, assembled into micelles, are found exclusively in milk. The macromolecular structure of native caseins makes them highly resistant to heat treatment. Their structure is also relatively well preserved when exposed to lipid or gas interfaces. However, they are sensitive to any change in the ionic environment and to proteolytic enzymes (rennet). Caseins and their derivatives (caseinates) are used in the food industry for their water retention capacity as well as their thickening, gelling and interfacial properties.

5.3.1.2. *Thickening and gelling properties of polymers*

Polysaccharides or proteins are widely used in the production and stabilization of food systems due to their thickening and gelling properties. The thickening and gelling properties depend primarily on the characteristics and behavior of the polymers in solution, especially their molecular weight and conformation, which affects the hydrodynamic volume and the degree of interaction between the polymers and solvent (usually water molecules). If the bonds between polymers are absent or weak, they can be easily separated, by mechanical treatment for example. Due to their interactions with the solvent, they exhibit thickening behavior. If, however, the bonds between certain polymer segments are strong enough and the polymer concentration is high enough, a three-dimensional network can be formed, entrapping the entire solvent phase. In this case, the polymer displays gelling properties.

Unlike proteins, the thickening and gelling properties of polysaccharides exist at very low concentrations in water, often well below 1%, except for some like starches (2 to 5%) or gum arabic (>40%).

Thickening properties

All water-soluble polymers increase the viscosity of solutions. However, some, like most proteins, precipitate before the desired viscosity level is reached. The thickening properties of polymers depend on their concentration and intrinsic viscosity $[\eta]_0$.

Intrinsic viscosity can easily be determined by viscosity measurements in dilute media. In dilute solutions, that is at a concentration below the critical overlap concentration of polymers, each molecule occupies its own space and can move in solution independently of others. As the concentration increases, the polymers become entangled, which is the critical overlap concentration. Beyond this concentration, the different polymers are entangled without strong interactions between them. The critical overlap concentration is an inverse function of the intrinsic viscosity of polymers; it varies greatly from one polymer to the next. For a given molecular weight, the intrinsic viscosity of a compact polymer is lower, and therefore the critical overlap concentration is higher than that of a highly disordered linear polymer.

From a rheological point of view, beyond the critical overlap concentration, there is an abrupt change in the relationship between the specific viscosity extrapolated to zero shear rate and concentration. This is due to the fact that polymers can only move in an entangled network by slipping between neighboring molecules. This results in non-Newtonian behavior, that is a variation of apparent viscosity with shear rate (Figure 5.7).

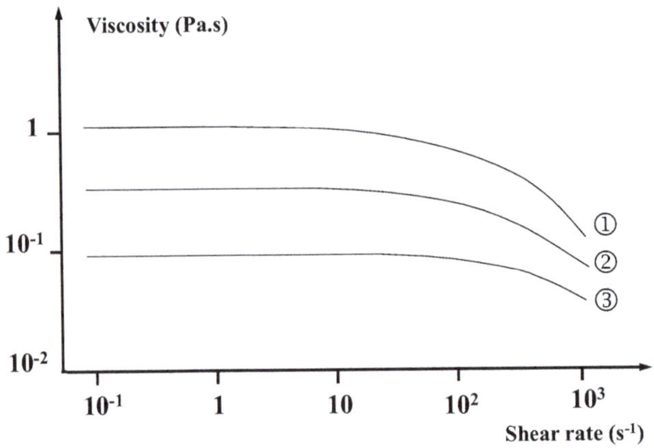

Figure 5.7. *Sample flow curve of an alginate solution (based on [DER 09]). Concentrations: ① 1.5 %, ② 1 % and ③ 0.7 %. Non-Newtonian behaviour occurs beyond a shear rate, which decreases as the concentration increases*

The intrinsic viscosity of polymers depends on their molecular weight, their conformation in solution and their flexibility. There are four main types of conformations:

– compact globular conformation;

– random coil conformation;

– rigid rod-like conformation;

– helical conformation.

Polymers with a compact globular conformation occupy a limited space even with a high molecular weight. Thus, they have a low intrinsic viscosity that is independent of their molecular weight. The random coil formation corresponds to a statistical distribution of the macromolecular chain in space. Almost all polymers can, under certain conditions, adopt a random coil formation. The rigid rod-like formation is a particular feature of the random-coil formation when the flexibility of the chain is restricted. This can be observed when polymers have charged groups (sulfate or carboxylate), which exert electrostatic forces leading to the stretching of the chain, or chain branches like in the case of xanthan gum. Finally, the helical conformation can be adopted only if stable intramolecular bonds allow the establishment of a helical structure. The intrinsic viscosity of these polymers depends on their average molecular weight according to Mark-Houwink type relationships:

$$[\eta]_0 = K\, M^{a} \qquad [5.7]$$

Parameters K and a depend on the given polymer/solvent/temperature system. For flexible polymers, parameter a generally varies from 0.5 to 1 whereas for rigid polymers it can reach 2. For the same molecular weight, chain branching usually results in the reduction of the intrinsic viscosity. Thus, branched polymers produce less viscous solutions. However, they also provide more stable solutions by limiting the possible overlapping areas between polymers.

Gelling properties

A gel can be defined as a system comprising at least two phases. It consists of a solid, three-dimensional macromolecular network that retains the entire liquid phase in its network. Although gels can be formed from proteins or polysaccharides, they have different characteristics:

– gels obtained from polysaccharides have a thin, transparent structure which forms at low concentration. This is the case for many gelled systems such as custards or jellies;

– protein gels are mostly opaque and are obtained at higher concentrations (5 to 10%). Gelation plays a major role in the preparation of many foods including various dairy products. Other examples are coagulated egg white, gelatin gels, various cooked products made from minced meat or fish, soy protein gels, textured vegetable protein by extrusion or spinning, and doughs.

Gelled systems are easily characterized by the use of rheological methods, especially in dynamic mode; for more information, refer to [BRU 06, MAL06].

To form a gel, polymers must firstly be dispersed in solution. Thus, prior to gelation, polymers form a true solution. The formation of a gel involves the linking of chains or chain segments to each other. These interactions can form naturally during cooling (gelatin gel, starch gel), after the unmasking of reactive groups (thermotropic globular protein gels), by adjusting pH (highly methylated pectin gels, acid casein gels, pre-denatured globular protein gels), by adding specific ions such as calcium (alginate or κ-carrageenan gel) or by the action of specific biological agents (rennet casein gel, fibrin gel). Sometimes, it is the mixture of different non-gelling polymers that allows the formation of a gel. Such mixtures are of high technological interest and are therefore used on an industrial scale. This is the case with xanthan-carob mixed gels. The reversibility (thermal, mechanical or ionic) of most gels used in the food industry suggests that the forces binding the chains together have enough freedom to allow change.

Several transitional stages can be identified during gel formation:

– the “sol” state where the polymer forms a solution: polymers are unordered in solution;

– the “gel” state occurs when enough chains link to form a network or elastic gel. The more ordered the chains are, the more rigid the gel becomes. Syneresis can subsequently occur in some cases (starch, casein or κ-carrageenan gels): the gel contracts and exudes a portion of the liquid phase.

Thus, the properties of a gel depends on the balance between interaction forces, which bind the polymer chains thereby avoiding flow, and repulsive forces (electrostatic, steric), which create pockets filled with the solvent phase.

5.3.1.3. *Interfacial properties of polymers*

Many food products are emulsions, foams or whipped emulsions (e.g. milk, cream, ice cream, butter, mayonnaise, finely minced meat like sausage meat, whipped egg white, bread, etc.). Polysaccharides (e.g. modified starches), but also proteins, often play an important role in the formation and stabilization of these colloidal systems. For example, a natural milk emulsion is stabilized by the fat globule membrane. It consists of successive adsorbed layers of mainly phospholipids and proteins (insoluble lipoproteins and soluble proteins). Homogenizing milk increases the stability of the emulsion by reducing the fat globule size and adsorbing proteins (casein micelles, soluble proteins) to the newly-formed fat globule surface. These macromolecules provide the interface with physicochemical and rheological properties (steric hindrance, electrostatic repulsion, viscoelasticity), which determines resistance to droplet coalescence. During the beating of egg whites, proteins adsorb on the newly-formed interface, denature and form a "gel" network that traps air bubbles and whose rheological properties govern the stability of the foam.

Several theoretical and experimental studies have been conducted on protein behavior at water/oil and water/air interfaces. However, uncertainty remains with regard to the conformation of proteins at interfaces, interactions between proteins at the interface and molecules of the underlying phase, as well as their impact on emulsifying and foaming properties. In order for a macromolecule to have good interfacial properties, it must:

– migrate and adsorb at the newly-created interface;

– rapidly form a cohesive interface layer ensuring dispersion stability.

Migration and adsorption of proteins at interfaces.

Due to their amphiphilic nature, proteins naturally adsorb at air/water or oil/water interfaces as their free energy is lower there than in either of the phases. However, in order for proteins to migrate to the interface and become surface-active, they must be soluble. A positive correlation often exists between protein solubility and the capacity to form foams or emulsions (soy protein, peanut protein, casein). Apart from their solubility, size and surface hydrophobicity of proteins, as well as the viscosity of the continuous phase, affect their diffusion and adsorption rates at interfaces. In the absence of external energy, an increase in the concentration of proteins at the interface is controlled by mass transfer, that is molecular diffusion. According to the Stokes–Einstein equation, the diffusion coefficient of a molecule is inversely proportional to the hydrodynamic radius and viscosity of the continuous phase. Thus, the homogenization of egg whites improves its foaming capacity while decreasing its viscosity.

Hydrophobic amino acids on the surface of proteins act as interface anchor points. A peptide segment of a few amino acids may suffice for adsorption if correctly located on the protein surface. The increase in the surface hydrophobicity of proteins reduces their affinity for the aqueous phase, and increases the probability of effective contact with a newly-created interface; they therefore migrate and adsorb more quickly at the interface. Appropriate treatment, such as moderate heat treatment, is sometimes used to increase the hydrophobicity of proteins and their interfacial properties. If the surface hydrophobicity of proteins becomes too high, protein solubility decreases, which reduces their capacity to form foams and emulsions.

Formation of cohesive interface layers and stabilization of dispersions

The contact between a protein and an interface generates asymmetry in its structural environment. The non-polar amino acid residues, initially mostly buried inside the protein structure, are then exposed to the hydrophobic phase (oil, air). This unfolding, the level of which depends on the given proteins and interface, results in a denatured protein layer that is strongly adsorbed at the interface, and leads to a reduction in the free energy of the system. Experimental studies have shown that the flexibility of proteins, their net charge, and the balance of intramolecular and intermolecular interactions between interface proteins are important parameters in the creation of a cohesive layer that stabilizes dispersions.

With emulsions, there is a strong correlation between emulsifying properties and the surface hydrophobicity of proteins. This suggests that proteins that adsorb at the lipid interface only undergo limited structural rearrangements. Caseinates are good emulsifiers because they have, in addition to high solubility, a naturally unordered structure (random coil), as well as relatively high overall hydrophobicity and a clear separation of hydrophilic and hydrophobic regions along the polypeptide chains. They form adsorbed protein films that are thick, highly hydrated and charged, allowing strong steric and electrostatic repulsions between lipid droplets and therefore ensure optimal emulsion stability. However, these emulsions can be sensitive to changes in pH or the addition of calcium through the modification of intramolecular and intermolecular interactions between proteins within the interfacial film and with proteins in the solution. Casein micelles and actomyosin (meat and fish proteins) also have good emulsifying properties. In contrast, globular proteins with a stable rigid structure, such as whey and egg white, proteins are not fully suitable as emulsifying agents since they do not unfold completely and are poorly anchored at lipid interfaces. An increase in hydrophobicity or molecular flexibility by a suitable pre-treatment (moderate heat treatment, partial proteolysis, disulphide reduction), without a loss of solubility, improves the emulsifying properties of globular proteins.

Apart from the nature of the emulsifying macromolecule used, many other factors influence the properties of emulsions, such as the type and dimensions of the homogenizer, the energy input to create emulsion, the oil addition rate, the volume of the lipid phase, temperature, pH, ionic strength, the presence of sugars, low molecular weight surfactants, exposure to oxygen and the nature of the oil (melting point).

Many proteins have a good ability to form foams. This is the case with egg white protein, globin and hemoglobin, serum albumin, gelatin, whey protein, casein micelles, β casein, wheat protein (glutenin), soy protein and some protein hydrolysates (low degree of hydrolysis).

However, all these proteins are not able to form a film that stabilizes the foam during storage. β casein and caseinates, flexible molecules with an unordered structure (random coil), quickly lower interfacial tension and facilitate foam formation. However, adsorbed protein films are thin and foam stability is poor.

In order to obtain a stable protein foam, the proteins used must be able to form a cohesive, continuous, viscoelastic, thick and air-tight film around each gas bubble. To create such a network, proteins that adsorb at the air/water interface must be:

– flexible enough to adsorb irreversibly at the interface and expose reactive groups to establish intermolecular interactions (cohesive and continuous film);

– rigid enough to preserve the secondary structures in the native protein (viscoelastic film);

– not highly charged (thick film);

– not highly hydrated (air-tight film).

For these reasons, globular proteins that remain soluble at a pH close to their isoelectric point generally form stable foams. They partially unfold at the air/water interface, which allows them to adsorb by multiplying the points of contact through hydrophobic interactions. This is necessary to avoid desorption of the proteins and their transport with the draining fluid along the interface. Meanwhile, these globular proteins expose reactive groups to form intermolecular associations (hydrophobic, electrostatic, disulphide) necessary for a continuous and cohesive adsorbed film, while preserving the secondary structure elements that give the film good rheological properties.

Stable interfacial films are obtained for a critical amount of cohesive intermolecular interactions between proteins. Excessive protein aggregation results in lamella rupture. A viscoelastic interfacial film and sufficient mobility of protein molecules are required to counteract the shear forces exerted on the film (expansion of the film that occurs at the top of the air bubbles and compression at the bottom). Any expansion of the film causes a decrease in the concentration of adsorbed molecules and an increase in interfacial tension. In order to stabilize foams, proteins should be able to migrate to the region of high interfacial tension to restore the initial thickness of the lamellae (Marangoni effect).

In addition, thick and less hydrated films stabilize the foam with respect to coalescence and Ostwald ripening. Electrostatic repulsions between proteins at the air/water interface decrease the cohesion of the film and increase the risk of rupture of bubble lamellae. Electrostatic repulsions are minimal at the isoelectric point of proteins, allowing the adsorption of greater amounts of proteins at the interface. Protein films are therefore thicker and foam stability is increased. Proteins also revealed minimal hydration at their isoelectric point, which limits the disproportion between bubbles of different sizes, with gas diffusion occurring in the aqueous phase.

5.3.2. *Low molecular weight emulsifiers*

Low molecular weight emulsifiers are widely used in the food industry. They are often needed to stabilize complex food systems involving immiscible phases. In recent years, the development of low-fat products with added water has led to a greater use of these emulsifying agents.

Emulsifiers are amphiphilic compounds whose chemical structure comprises both hydrophilic and hydrophobic functions. This particular chemical structure gives them the ability to adsorb at oil/water interfaces and thus ensure foam stability. The amphiphilic structure of these molecules, however, gives them unique physical properties allowing them to stabilize air/water interfaces (foams), form complexes with polymers (proteins and polysaccharides) or liposomes, and control the crystallization of fats. These characteristics have a significant impact on sensory properties and the physical stability of food during storage.

5.3.2.1. *Origin and chemical structure*

Monoglycerides and their synthetic derivatives are the most common emulsifying agents. The rest of the market is dominated primarily by lecithin, a natural by-product from the oils and fats industry. Other molecules have been developed by chemical or enzymatic synthesis with the aim of obtaining more effective products. This is the case, for example, with sucrose esters, but their use in food is marginal due to their high cost.

Lecithin, which is also marketed as a nutritional supplement because of its high level of polyunsaturated essential fatty acids and its cholesterol-lowering effect, was not such a commercially successful emulsifier as expected mainly due to its limited surfactant properties for some food applications. Moreover, the opportunities resulting from the fractionation and/or modification of these natural molecules creates more effective, but also more expensive products. Finally, the image of lecithin has recently been tainted following the appearance of genetically modified soybeans, which are a major source of lecithin (Table 5.3).

Type	*Origin and structure*	*Physicochemical properties and features*	*Application*
Monoglycerides & derivatives of monoglycerides (E471-E472)	Transesterification reaction between a triglyceride and glycerol at high temperature (200 to 250°C), under vacuum and in the presence of a catalyst (usually sodium hydroxide). Mixture containing 40 to 60% monoesters, 30 to 40% diesters and 10 to 20% triesters.	Main category of food emulsifiers. Only monoesters have desired surfactant properties.	Derivatives obtained from organic acids and ethylene oxide, used in English-style bread making to improve crumb texture and loaf volume. Citroglycerides used in margarine to improve cooking properties and reduce spattering during frying. Acetoglycerides widely used to stabilise foams. Polyethylene oxide derivatives (very hydrophilic) used as conditioning agent in bread dough and as an emulsifier in ice cream and whipped cream.
Sucrose esters (E473) and sucroglycerides (E474)	Sucrose esters: fatty acid esters obtained by direct esterification of sucrose by fatty acid methyl esters.	Easy dispersibility in water. Mild taste.	Diverse applications: ice cream, baking products, margarine, emulsified sauces, etc.

	Sucroglycerides: mixtures of partial glycerides and sucrose esters obtained by the transesterification of sucrose and triglycerides in a solvent medium.		High price/property ratio compared to monoglycerides, even if produced by chemical synthesis: limited use in food.
Sorbitol esters (E491-5)	Esterification of one or more hydroxyl groups of sorbitol (obtained by hydrogenation of glucose) and sorbitan (obtained by dehydration of sorbitol) with fatty acids (mainly stearic and oleic acid).	Better known under the trade name "SPAN" Polyoxyethylene derivatives ("TWEEN") with hydrophilic properties. Alter crystallinity of fats and delay crystalline transition to a more stable form.	Wide range of HLB (hydrophilic – lipophilic balance; equation [5.8]) allows multiple applications such as water/oil and oil/water emulsifier in the food industry. Crystal stabiliser for cocoa butter to prevent migration of fat from the inside to the surface of the chocolate ("anti-blooming agent").
Phospholipids (lecithins) (E322)	Egg yolk lecithin: unprofitable (high extraction cost). Commercial lecithin obtained by degumming: decrease in the solubility of phospholipids in soybean oil in the presence of 1 to 3% water and 1 to 3‰ phosphoric acid. Complex mixtures of phospholipids. Glycolipid fraction composed of glycosylated sterols, mono and digalactosyl diglycerides and complex glycolipids. The fatty acid composition (palmitic, oleic, linoleic and linolenic acid) characterizes each phospholipid class.	Relatively limited emulsifying properties: -phosphatidylcholine: stabilisation of oil/water emulsions. -phosphatidylethan-olamine and phosphatidylinositol: water/oil emulsifying properties. Purified fractions: improved emulsifying properties but high extra cost.	Natural phospholipids have a large variety of food applications: bakery products, chocolate products, instant products, margarine, etc. In the case of chocolate, the addition of lecithin can modify the rheological characteristics of melted chocolate and thus save on other expensive fats (cocoa butter). Phospholipids are known for their antioxidant properties that result from the presence of traces of tocopherols in the oil medium.

Table 5.3. *Main emulsifiers and their food applications*

5.3.2.2. *Properties and functions of emulsifiers*

Physicochemical properties (solubility, mesomorphism)

The effectiveness of an emulsifier is primarily related to its solubility in both phases (water and oil). The solubility of any surfactant is characterized by its hydrophilic-lipophilic balance (HLB). This index, based on semi-empirical considerations, is used to estimate the hydrophilicity of the emulsifying agent by taking into account the relative proportions of hydrophilic and hydrophobic parts. This parameter is defined by:

$$\text{HLB} = \frac{\text{molecular mass of hydrophilic portion}}{\text{molecular mass of the whole molecule}} \times 20 \qquad [5.8]$$

Theoretically, the HLB values of each emulsifier range between 1 and 20. The lower this value is, the more lipophilic the emulsifier and vice versa. Generally, emulsifiers with a HLB between 1 and 6 stabilize water/oil emulsions whereas those with a HLB between 8 and 18 stabilize oil/water emulsions. It is also possible to estimate the HLB of a mixture of emulsifiers by calculating the arithmetic average of the HLB of each constituent. Knowing this index can help to choose which emulsifier to use, but, in practice, it cannot be used to predict the effectiveness of each emulsifier or emulsifier mixture at stabilizing a given emulsion (synergistic or antagonistic effects). Thus, two emulsifiers with the same HLB will not necessarily have the same functional properties. Measuring the hydration of the polar head of emulsifiers would optimize the properties of emulsifier mixtures under their conditions of use (temperature, ionic force, pH, etc.). Indeed, hydration of the polar head is a factor that contributes significantly to the organization and rheological properties of interfacial films.

The effectiveness of an emulsifying agent in the formulation of a food product is determined by its physical state, which depends on chemical structure and two key parameters: temperature and water content, both of which determine lipid chain packing. Thus, for each type of emulsifier, there are different characteristic structures called mesophases. The different types of mesophases observed with different emulsifying agents are micellar, lamellar and cubic phases. This polymorphism has been described in great detail for a large number of food emulsifiers in [KRO 76] and summarized in [DER 09].

Transfer of surfactants to interfaces

The diffusion of amphiphilic molecules dispersed in the aqueous phase to the interface and the formation of a monomolecular film are the basic mechanisms for the expression of surfactant properties. When molecules form micelles in aqueous solution, this adsorption occurs rapidly due to an equilibrium between molecules in solution (monomers) and aggregated molecules (micelles) [DIC 92]. When surfactants form lamellar phases (e.g. lecithin), adsorption kinetics is slower.

Stabilization of emulsions

This fundamental property is based on the amphiphilic nature of the emulsifier. It involves stabilizing a system composed of two or more immiscible phases. The emulsifier is adsorbed at the oil/water interface, which has the effect of lowering the interfacial tension. The rheological properties of the interfacial film depend on the molecular arrangement of the emulsifier. If it is not very concentrated, the molecules are oriented parallel to the interface and are not in an ordered state ("liquid") although some cohesive forces occur between the chains. At higher concentrations, the films are more condensed; the molecules are stacked and positioned perpendicularly to the film surface with polar groups facing towards the aqueous phase and hydrocarbons chains towards the lipid phase. Molecules cannot move due to steric hindrance and film cohesion is enhanced.

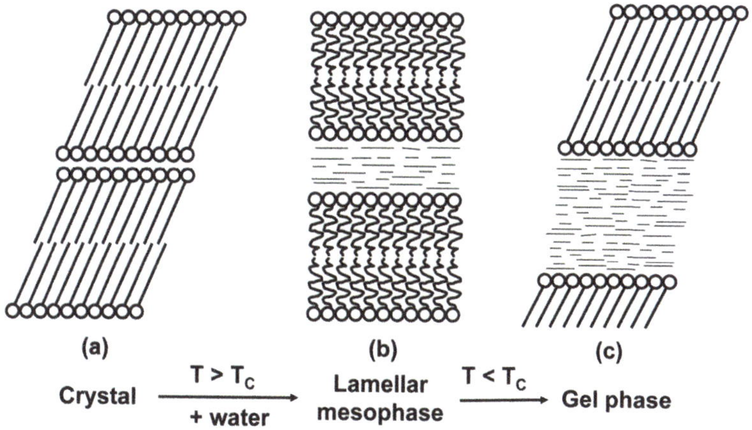

Figure 5.8. *Schematic representation of a surfactant transition from a crystalline structure a) to a fluid liquid crystalline lamellar mesophase b) and liquid crystalline gel c) (based on [KRO 76])*

Generally, in an oil/water emulsion, a lamellar structure will form at the interface (Figure 5.8) whereas in a water/oil emulsion it will most certainly be a cubic or hexagonal II (H_{II}) structure (Figure 5.9).

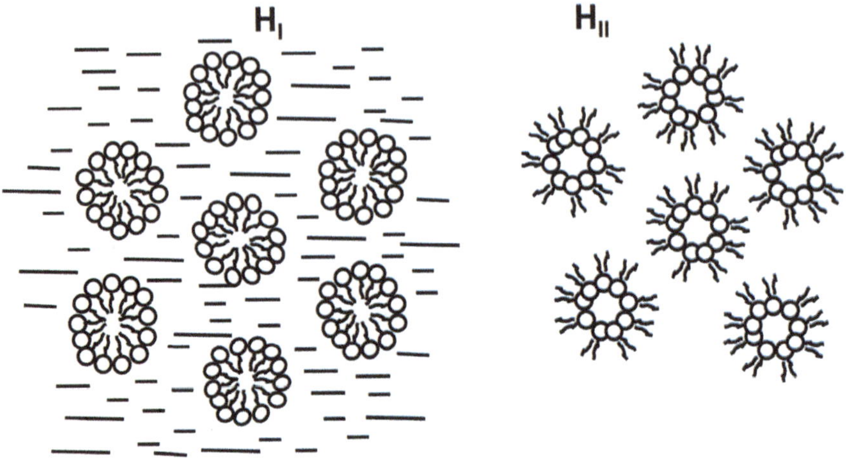

Figure 5.9. *Schematic representation of normal (micellar) H_I and inverted H_{II} hexagonal phases (cross section of cylindrical stacks)*

The existence of these liquid crystal structures explains why saturated or unsaturated monoglycerides are able to stabilize oil/water emulsions despite their low HLB. If the emulsifier concentration is sufficient, lamellar multilayers can form at the interface and stabilize this type of emulsion.

Stabilization of foams and whipped emulsions

The mode of action of emulsifiers in "aerated" systems is also linked to their mesomorphism. It generally involves complex emulsion systems of mainly fats and proteins. In the case of aerated emulsions (e.g. whipped cream), the function of the emulsifier is to help give volume, texture and stability against syneresis. The cohesion of the aerated emulsion is ensured by partially coalesced fat globules forming a thick network inside the foam lamellae. These are α-crystalline emulsifiers (propylene glycol monostearate, acetylated or lactylated monoglycerides) that are the most effective in promoting partial coalescence of fat globules.

These emulsifiers do not exhibit polymorphism and can only exist, below their melting point, as α-crystalline (Figure 5.10). They are easily soluble in fats and form a rigid crystalline film on the surface of fat globules. The mode of action of these emulsifiers is explained schematically by the following hypothesis: in the absence of emulsifier, proteins are adsorbed at the oil/water interface and prevent the partial coalescence of fat globules; the presence of an emulsifier leads to their adsorption at the interface and the displacement of proteins into the aqueous medium; protein displacement is facilitated by decreasing temperature. By removing proteins, emulsifiers decrease oil/water interfacial tension, but also the cohesion of the fat globule membrane, promoting the aggregation of fat globules [KRO 77]. This results in a slight decrease in foam volume but an increase in foam rigidity.

This mechanism helps to explain the role of more polar emulsifiers or monoglycerides in the preparation of ice cream. During maturation of the homogenized emulsion at low temperature, fat crystallizes and proteins adsorbed at the surface of fat globules are displaced by low molecular weight emulsifiers. The membrane surrounding the fat globules is thus weakened. During collisions throughout the whipping stage, the fat crystals formed perforate the weakened membranes and cause the partial coalescence of fat globules. This also occurs in pastry containing fat where foaming is generally obtained using egg proteins, but is hindered by liquid fat that destabilizes the foam structure. The incorporation of α-crystalline emulsifiers limits this effect by preventing oil droplets from coming into contact with proteins in the aqueous phase.

Moreover, in the case of low-fat cakes, the rising of the dough can be improved by the incorporation of saturated monoglycerides. This incorporation leads to a more uniform distribution of air in the dough, proper rising and satisfactory cake volume. The mode of action of monoglycerides is probably due to the formation of a monoglyceride/protein film at the air/water interface, and this film makes the foam more stable against coalescence [SHE 76].

PART 4

Food Ingredient Preparation

6

Physicochemical Basis of Fractionation and Related Technologies

The principle of extracting a compound or group of compounds is based on their physicochemical properties or specific features compared to the other components in the medium. These features generally include the molecular or supramolecular structure, whether native or induced by physicochemical changes (pH, ionic strength, dielectric constant and temperature), the hydrophilic/lipophilic nature of molecules, their ionic properties and their specific affinity to a particular ligand (enzyme–substrate, antibody–antigen). Various physicochemical separation techniques have been developed in recent years, which have contributed to the emergence of the food ingredients sector. Current industrial-scale processes include cell and particle separation (centrifugation, microfiltration and sedimentation) and component fractionation based on steric differences (ultrafiltration and gel permeation chromatography), ionic properties (ion exchange and electrodialysis), lipophilicity (organic solvent) and the specific affinity to a ligand (affinity chromatography).

The extraction processes always generate by-products that may be potential pollutants and for which food or non-food uses must be found. The valorization of by-products should also be taken into account in the techno-economic evaluation of separation processes, even if they are applied to the production of high value-added products.

Chapter written by Romain JEANTET.

6.1. Particle separation

The fraction to isolate or enrich may be dispersed (emulsion, colloidal suspension) or solubilized in an aqueous or lipid phase; the water soluble or fat soluble molecule is dependent on the physicochemical conditions of the medium (temperature, pH, ionic strength, ionic environment, dielectric constant, etc.) and can be altered by technological treatments such as heat treatment. Treatments that induce the precipitation/crystallization of fractions without impairing their nutritional and functional properties or the quality of the dispersing phase are generally the preferred option for processing due to the moderate cost and high control levels throughout the food sector.

6.1.1. *Aggregation, precipitation and crystallization of molecular elements*

6.1.1.1. *Isoelectric, ionic and heat-induced aggregation*

The solubility of hydrophilic molecules depends on the type of their functional groups, molecular weight and structure. Functional groups with a high affinity to water include "hydroxyl" groups, which are abundant among carbohydrates, and cationic (NH_3^+) and anionic (CO_2^-, SO_3^-, PO_4^{2-}) groups that play a key role in the solubility of proteins and some polysaccharides. Hydrophobic groups include linear or cyclic "alkyls" such as those in the side chains of triglycerides and some amino acids (leucine, isoleucine, proline, phenylalanine and tryptophan). Some molecules have both hydrophobic and hydrophilic groups (amphiphilic), for example fatty acids (C_n H_p CO_2^-), phospholipids, mono- or diglycerides or certain proteins. These molecules are able to position themselves at water/lipid or water/air interfaces or to structure themselves so that the hydrophobic sites are exposed as little as possible to the aqueous phase, which can result in the formation of a micelle or liposome structure with a hydrophobic core and a hydrophilic surface. Intermolecular interactions by hydrogen, van der Waals or ionic bonding can increase molecular size and limit solubility. These interactions occur less when the molecules have a high charge density, generating electrostatic repulsion.

The solubility of molecules with acidic or basic ionizable groups is highly dependent on the physicochemical conditions of the medium (pH and ionic environment; see Volume 1 [JEA 16a]). In the case of acids, charge and solubility increase as the pH rises (pH > pK_A) and in the case of amines, protonation and solubility increase as the pH drops (pH < pK_B). When molecules have both types of ionizable groups (amphoteric), which is the case with proteins, the overall charge is positive at low pH, negative at high pH and zero at intermediate pH, known as the isoelectric point (pI). At the pI, proteins have a minimum of hydration and electrostatic repulsion, which results in a minimum of solubility as shown in Figure 6.1.

The pI of amino acids and proteins depends on the pH at which half the acid groups are deprotonated and half the basic groups are protonated (respectively equivalent to the pK_A and pK_B at weak ionic strength). This point can vary with the ionic strength of the medium. The apparent pK or the pH at which half of the acid groups are deprotonated decreases with ionic strength, and that of basic groups increases. As a result, the pI of acidic proteins can decrease and that of basic proteins can increase, for example the pI of milk casein can vary by 0.5 pH units for an ionic strength between 0 and 0.1 M.

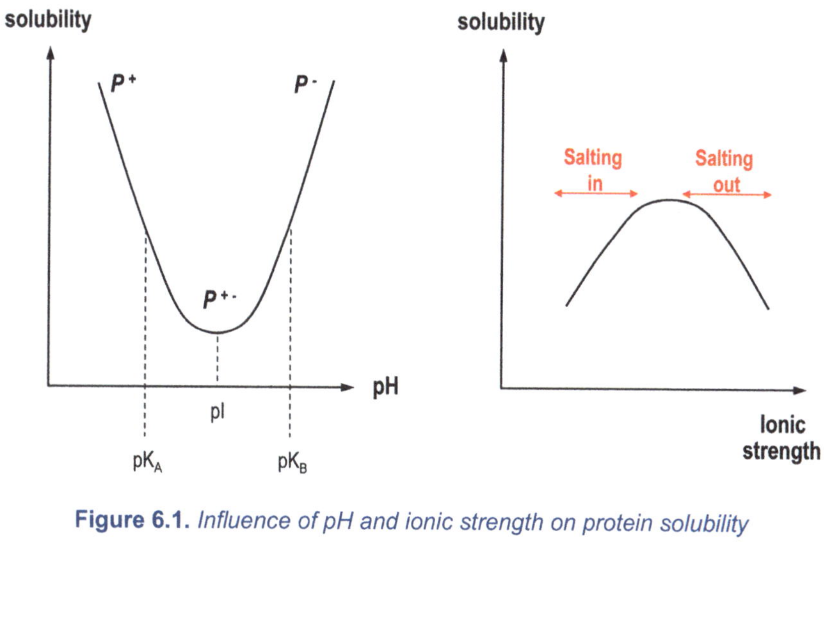

Figure 6.1. *Influence of pH and ionic strength on protein solubility*

Protein solubility can also depend on the ionic strength of the medium. Precipitation can be achieved by either reducing or increasing ionic strength. The aggregation and precipitation of some polysaccharide and protein compounds can be induced by ions that associated with them, resulting in charge screening, and/or by ions that change the properties of water and favor solvent properties (salting in) or disfavor them (salting out). Precipitation by salts has the major drawback of increasing the mineral content of by-products and thus complicating their valorization or further processing.

The aggregation or precipitation of protein compounds can often be promoted by heat treatment. Some proteins are stable over a wide pH range by their tertiary structure exposing hydrophilic groups on the outside and masking hydrophobic groups on the inside. Heating generally alters the tertiary structure and sometimes the secondary structure, which results in the unmasking of hydrophobic groups and/or highly reactive groups such as SH groups and favors hydrophobic and covalent interactions by SH-SS exchanges. Heat-denatured proteins precipitate during treatment or become very sensitive to pH and ionic strength. The level of protein denaturation and aggregation, and consequently solubility, depends on the protein concentration, the reaction order, the physicochemical properties of the medium (pH, ionic strength, type of ions) and the heat treatment conditions (temperature and temperature profile).

6.1.1.2. *Precipitation by lowering the dielectric constant of the solvent*

Ionic interactions (ion–ion, ion–dipole), hydrogen bonds (dipole–dipole) and hydrophobic interactions play an important role in the structure of protein and carbohydrate compounds and their ability to bind water, which determines their solubility. The dielectric constant of the solvent is an important parameter upon which the potential energy of ionic interactions depends. As the dielectric constant of water is high, the energy of ionic bonds is weak and electrostatic repulsion forces prevail, resulting in good solvent properties of the ionic solutes. A reduction in the dielectric constant by increasing temperature or by adding a solvent such as alcohol therefore alters the structure and properties of compounds (predominance of attractive forces) and weakens the ability of the solvent to solubilize ionic species.

Salts and some protein and carbohydrate compounds can thus be precipitated in a water-alcohol medium.

6.1.1.3. *Crystallization*

Crystallization is a change in state from a solute in solution to its solid state (sugar, salt) or from a molten to a solid state (triglycerides).

Highly hydrophilic solutes (salts, carbohydrates) have a high ability to bind water and can create limiting solvent conditions above a certain concentration known as the solubility threshold, beyond which non-dissolvable molecules remain in the solid state, in amorphous or crystalline form. The solubility threshold of sugars and salts generally increases with temperature. There are however exceptions, known as inverse solubility salts (e.g. calcium carbonate and calcium phosphate).

The crystallization of a solute implies that it reaches its solubility threshold and stays above it. In practice, the product is first concentrated (water is removed by evaporation or reverse osmosis) then gradually cooled to lower the solubility threshold and promote phase transition. The formation of small crystals or an amorphous solid is favored by very rapid cooling; otherwise, large crystals are formed, which can facilitate their extraction. Crystallization can also be initiated by the addition of small crystals to the concentrate (seeding).

To pass from a molten to a solid state, the temperature must be lowered to a value below the melting temperature. This is the case with triglycerides of oil or fat whose melting points can span a wide temperature range. Their melting point depends on the length of three fatty acid chains as well as their degree of unsaturation (number of double bonds). By gradually cooling the oil, triglycerides with a high melting point crystallize.

The phase transition is accompanied by a release of energy and a decrease in the concentration of solutes at the solid-liquid interface, which can slow down crystallization. It is then necessary to dissipate the energy and promote the transfer of solutes from the medium to the interface by moderate agitation to avoid breaking the forming crystals.

6.1.2. *Separation processes*

6.1.2.1. *Sedimentation and centrifugation*

The separation of particles whose density, ρ_p, is different to that of the dispersing phase, ρ_d, can be achieved naturally by sedimentation ($\rho_p > \rho_d$) or creaming ($\rho_p < \rho_d$). The sedimentation velocity v_s depends on the properties of the particles (ρ_p, diameter D) and those of the dispersing phase (ρ_d, viscosity η); it is expressed by Stokes' law [3.4]:

$$v_s = \frac{D^2}{18\eta}(\rho_p - \rho_d)\, g \qquad [6.1]$$

where g is the acceleration of gravity (81 m s^{-2}). Sedimentation can be carried out under batch or continuous conditions (Figure 6.2). In the case of a continuous settling tank, the sedimentation time t_s should be less than or equal to the residence time, t_r of the fluid carrying the particle, which is separated before removal through the bottom of the settling tank. If h and L are respectively the height and length of the settling tank and v_f the flow velocity of the fluid, it can be written as:

$$t_s = \frac{h}{v_s} \leq t_r = \frac{L}{v_f} \qquad [6.2]$$

The maximum flow velocity is therefore:

$$v_f \leq \frac{L}{h} v_s \qquad [6.3]$$

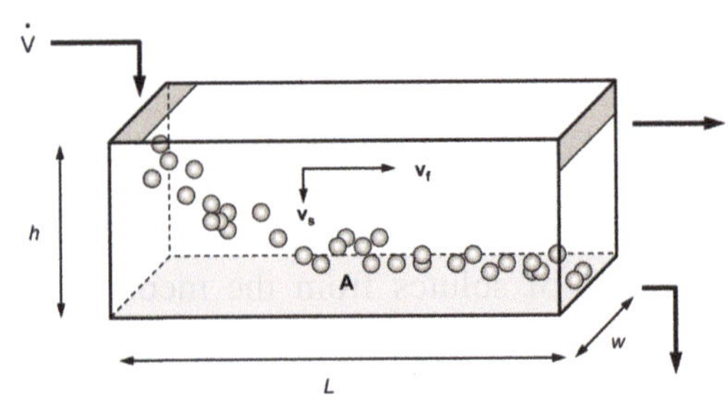

Figure 6.2. *Principle of a continuous settling tank*

If w is the width of the settling tank, the volumetric flow rate of the settling tank $\dot{V}$ is written as:

$$\dot{V} = h\, w\, v_f \qquad [6.4]$$

The flow rate limit for the separation of the particle is therefore ([6.3] and [6.4]):

$$\dot{V} = L\, w\, v_s = A\, v_s \qquad [6.5]$$

where A is the horizontal surface area of the settling tank (m^2). As a result, the flow rate limit is independent of the height of the settling tank: when the sedimentation path decreases (drop in h), the flow velocity increases by the same proportion (decrease in the flow section).

Particle separation can be accelerated by centrifugation; the diagram of a centrifugal separator is shown in Figure 3.2. The rate of separation depends on the properties of the particles and the dispersing phase as well as the centrifugation conditions (angular velocity ω and the centrifugal radius R; [3.5]):

$$v_s = \frac{D^2}{18\eta}(\rho_p - \rho_d)\,\omega^2 R \qquad [6.6]$$

6.1.2.2. *Microfiltration*

Particles can be separated by continuous or batchwise dead-end (Figure 6.3) or cross-flow filtration (Figure 6.4). In dead-end filtration, the particles accumulate on the surface of the filter and form a deposit, thereby increasing transfer resistance R. The volumetric flow rate $\dot{V}$ ($m^3\, s^{-1}$) depends on the pressure difference on either side of the filter surface (ΔP, Pa), the viscosity of the dispersing phase (η, Pa s), the hydraulic resistance (R, m^{-1}) and the filter surface (A, m^2). Darcy's law gives the permeation flux density J ($m.s^{-1}$), which is a ratio of $\dot{V}$ and A [3.8]:

$$J = \frac{\dot{V}}{A} = \frac{dV}{Adt} = \frac{1}{R}\frac{\Delta P}{\eta} \qquad [6.7]$$

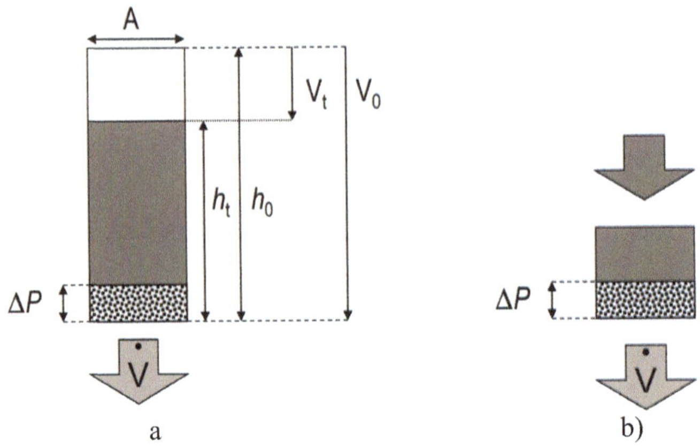

Figure 6.3. *Principle of dead-end filtration; a) static filtration; b) continuous filtration*

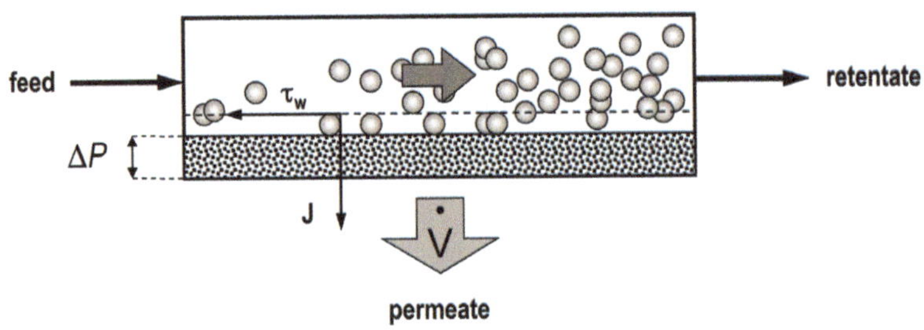

Figure 6.4. *Cross-flow filtration*

In the case of static filtration, the pressure difference at time t is equal to:

$$\Delta P = h_t \ \rho \ g = \frac{V_0 - V_t}{A} \rho \ g \qquad [6.8]$$

By combining [6.7] and [6.8] we get:

$$\frac{dV}{V_0 - V_t} = \frac{\rho \ g}{\eta \ R} dt$$

which integrates to:

$$V_t = V_0 \left(1 - e^{-\frac{\rho g}{\eta R} t}\right) \qquad [6.9]$$

One of the drawbacks of dead-end filtration is the increase in hydrodynamic resistance R as the particles accumulate on the membrane surface. If the pressure is increased to compensate for the increase in R, compaction of the particle deposit can increase R.

Cross-flow filtration has the advantage of limiting the accumulation of particles on the membrane surface. The transfer laws of cross-flow microfiltration are described in Chapter 3, section 3.2.

6.2. Steric separation

6.2.1. *Decreasing or increasing molecular size*

When the molecule to be isolated has a very different molecular weight to that of other components in the medium, molecular separation based on size is relatively easy: this is case for example with the separation of milk proteins, whey or egg albumen from carbohydrates and mineral salts. There is a large variation in molecular weight between these different components from less than 400 g mol^{-1} for all carbohydrates/salts to more than 15,000 g mol^{-1} for proteins.

In the case where molecules have a similar molecular weight, other separation techniques are required, which involve increasing the difference in molecular weight. This can be achieved by reducing the molecular weight of the compound to be removed by degradation or hydrolysis. Alternatively, the molecular weight of the compound to be extracted can be increased by promoting aggregation. For instance, in a protein/polysaccharide mixture, only the polysaccharides can be enzymatically hydrolyzed (by amylases or pectinases) if the proteins are isolated, or conversely the proteins can be enzymatically hydrolyzed (by proteases) if the polysaccharides are isolated. In some cases, it is possible to increase molecular weight through the molecular interaction of compounds to be isolated without reaching the solubility limit, for example the aggregation of phosphorylated peptides in a milk protein hydrolyzate using calcium ions results in high molecular weight aggregates.

6.2.2. *Separation processes*

6.2.2.1. *Ultrafiltration*

Cross-flow ultrafiltration is a widely used technique for molecular separation. It has undergone significant development over the past thirty years, especially in the food industry and its main applications are the concentration and purification of proteins.

Theory of solvent transfer

The laws of microfiltration (see Chapter 3, section 3.2) also apply to ultrafiltration, and therefore it is possible to express the permeation flux density J according to Darcy's law. In general, this model is refined by breaking down the overall resistance to solvent flow R into its components:

$$R = R_m + R_p + R_f \qquad [6.10]$$

where R_m, R_p and R_f (m^{-1}) are respectively the hydraulic resistances of the membrane itself, of the accumulation of compounds close to the membrane (polarization concentration; Figure 3.5) and of the fouling layer. Combining [6.7] and [6.10] results in the "hydraulic resistance in series" law (see Chapter 3, [3.9]):

$$J = \frac{1}{(R_m + R_p + R_f)} \frac{TMP}{\eta} \qquad [6.11]$$

R_p resistance linked to the accumulation of compounds on the membrane surface depends on the ultrafiltration parameters and especially the J/τ_w ratio, with τ_w being the wall shear stress shown in Figure 6.4. This ratio results in competition between the convective supply of molecules to the membrane which depends on J, and the removal of the deposit by erosion which increases with τ_w. As already mentioned, it is possible to temporarily compensate for the decrease in J due to arise in R by increasing the transmembrane pressure *TMP*; there is a risk however that the selectivity of the filter support is altered by compaction of the deposit, usually resulting in intensified fouling. R_p resistance increases with the concentration of compounds and R_f resistance with operating time.

Ultrafiltration can be carried out batchwise or continuously (Figure 6.5). In batch filtration, the flux density decreases with the volume reduction ratio (VRR), which is defined as the ratio between the initial volume (V_0) and the volume at time t (V_t) of filtered liquid, reflecting an increase in the concentration of compounds (Figure 6.6):

$$\text{VRR} = \frac{V_0}{V_t} \qquad [6.12]$$

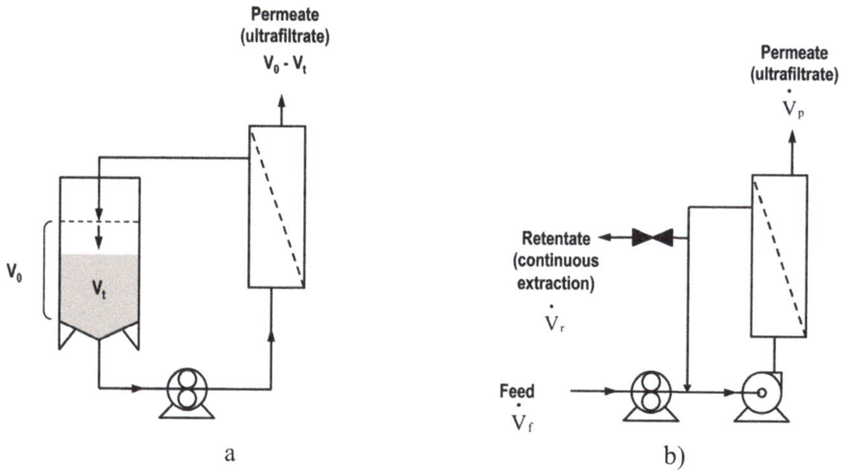

Figure 6.5. *Principle of ultrafiltration. a) Batch. b) Continuous*

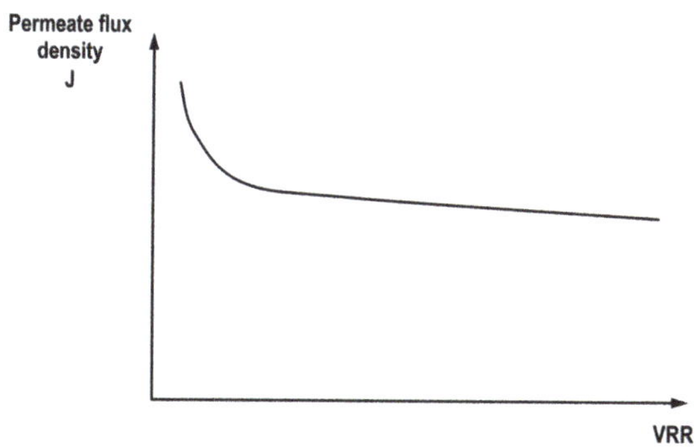

Figure 6.6. *Change in the permeation flux density J as a function of the volume reduction ratio*

In continuous filtration, the ultrafiltration membrane is fed at flow rate $\dot{V}_f$, which is equal to the sum of the flow rate of the permeate $\dot{V}_p$ and the retentate $\dot{V}_r$:

$$\dot{V}_f = \dot{V}_r + \dot{V}_p \qquad [6.13]$$

Filtration is performed at a volume reduction factor VRR:

$$VRR = \frac{\dot{V}_f}{\dot{V}_r} = 1 + \frac{\dot{V}_p}{\dot{V}_r} \qquad [6.14]$$

Filtration performance decreases as the operation progresses, that is as both the VRR and compound concentration increase. This is why it is preferable to divide a system into several stages.

Determination of membrane surface area

To improve performance, most industrial plants are divided into several stages with each stage working continuously at a defined volume reduction ratio, as shown in Figure 6.7. In this configuration, it is possible to identify the optimal operating conditions for each stage.

In this context, it is possible to determine the surface area at each stage based on the experimental permeation flux density J as a function of the volume reduction ratio VRR.

Let:

– VRR_1, VRR_2,…,VRR_n be the volume reduction ratios corresponding to each stage;

– A_1, A_2,…, A_n the exchange surface area at each stage (m^2), and

– J_1, J_2,..., J_n the permeation flux densities measured at each stage ($l\ h^{-1}m^2$).

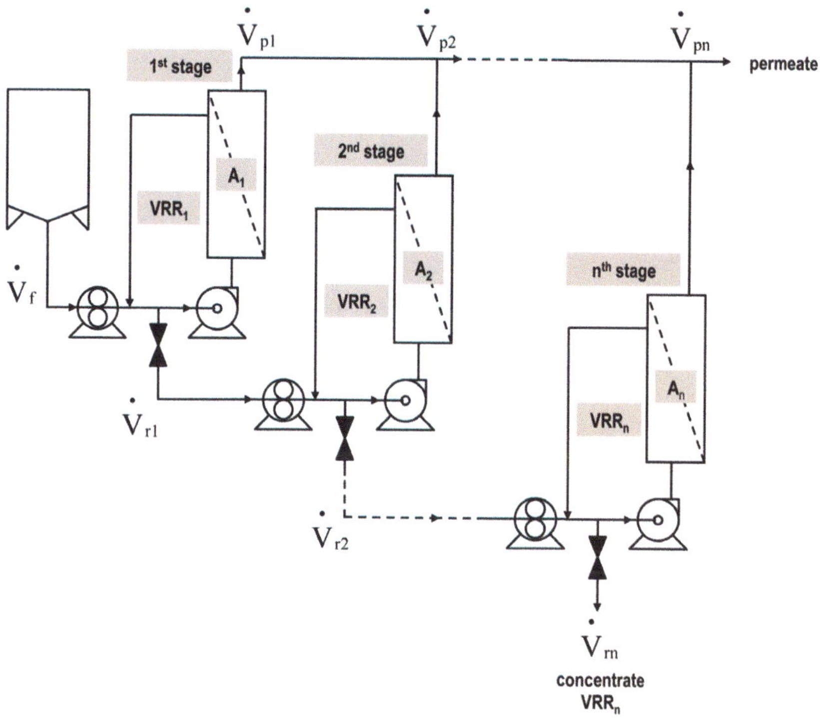

Figure 6.7. *Multistage filtration plant*

If $\dot{V}_f$, $\dot{V}_{r1}$ and $\dot{V}_{p1}$ are respectively the volumetric flow rates of the installation, the retentate and the permeate of the first stage, we can write:

$$VRR_1 = \frac{\dot{V}_f}{\dot{V}_{r1}} = \frac{\dot{V}_{r1} + \dot{V}_{p1}}{\dot{V}_{r1}} = 1 + \frac{A_1 J_1}{\dot{V}_{r1}} \quad [6.15]$$

Knowing that $\dot{V}_{r1} = \frac{\dot{V}_1}{F_1}$, the above equation can be written as:

$$VRR_1 = 1 + \frac{A_1 J_1}{\dot{V}_1} VRR_1 \quad [6.16]$$

which results in:

$$A_1 = \left(1 - \frac{1}{F_1}\right) \frac{\dot{V}_f}{J_1} \quad [6.17]$$

Similarly:

$$VRR_2 = \frac{\dot{V}_f}{\dot{V}_{r2}} = 1 + \frac{A_1 J_1 + A_2 J_2}{\dot{V}_{r2}} = 1 + \frac{A_1 J_1 + A_2 J_2}{\dot{V}_1} VRR_2 \quad [6.18]$$

which results in:

$$A_2 = \left(\frac{1}{F_1} - \frac{1}{F_2}\right) \frac{\dot{V}_f}{J_2} \quad [6.19]$$

Thus for each stage, the surface area is:

$$A_n = \left(\frac{1}{F_{n-1}} - \frac{1}{F_n}\right) \frac{\dot{V}_f}{J_n} \quad [6.20]$$

Knowing the relationship between $J = f_1(VRR)$, thus between $\frac{1}{J} = f_2\left(\frac{1}{VRR}\right)$, the optimum operating VRR for each stage can be plotted at a given feed rate $\dot{V}_f$ with respect to the total surface area installed. For example, in the case of a four-stage operation with volume reduction factors VRR_1, VRR_2, VRR_3 and VRR_4, the corresponding surface areas are indicated by the area of the rectangles in Figure 6.8. Preparing a concentrate at VRR_4 in a single-stage continuous system would, according to [6.20], lead to surface area A which is around three times larger than a 4 stage system:

$$A = \left(1 - \frac{1}{F_4}\right) \frac{\dot{V}_f}{J_4}$$

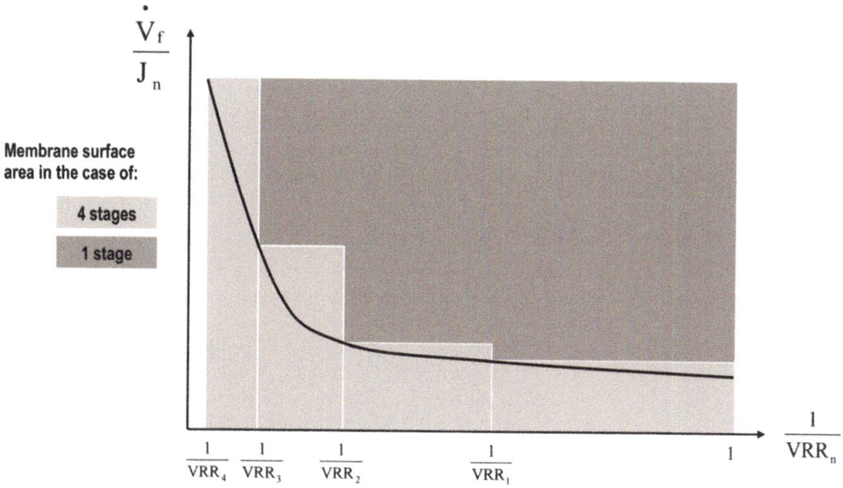

Figure 6.8. *Graphical representation of membrane surfaces based on the number of stages in the system*

The system can be scaled up by using the smallest membrane surface area, which would require the maximum number of stages, especially at high volume reduction ratios. However, above a certain number of stages, the cost of equipment (pumps, sensors, etc.) exceeds profit. The compromise is generally three to four stages depending on the applications (physicochemical properties of the products) and costs (membranes, environment and energy).

Diafiltration

Diafiltration is often required to obtain high-purity compounds. This involves adding water to the retentate, which dilutes the solutes that are not retained by the membrane and lowers the concentration of these later in the retentate (Figure 6.9).

To simplify the calculation, let us assume no further concentration occurs at the stage where diafiltration takes place: in this case, the volumetric flow rate $\dot{V}_r$ of this stage is equal to that of the diafiltered retentate ($\dot{V}_{dr}$) and the volumetric flow rate of water $\dot{V}_w$ to that of the permeate ($\dot{V}_p$):

$$\begin{cases} \dot{V}_r = \dot{V}_{dr} \\ \dot{V}_p = \dot{V}_w \end{cases} \quad [6.21]$$

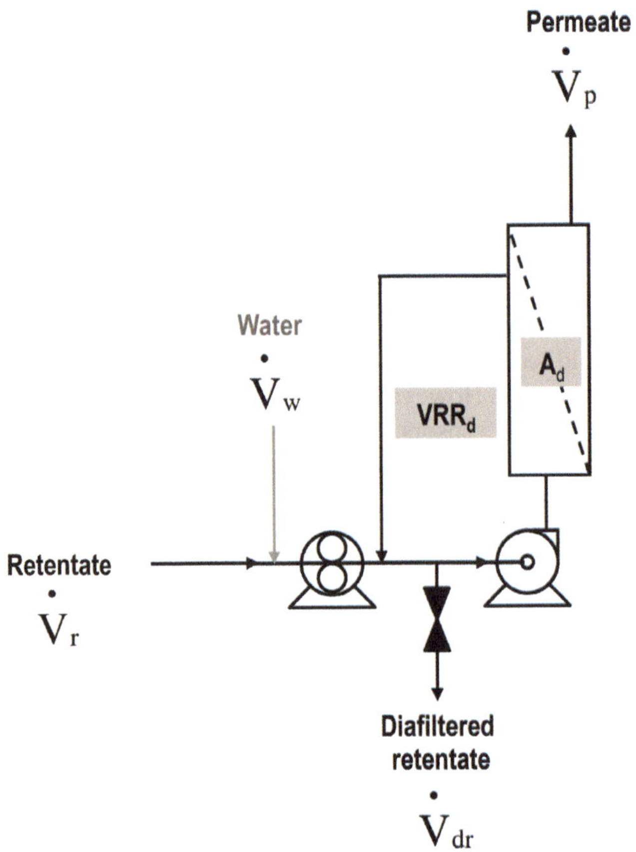

Figure 6.9. *Diafiltration system*

Given that:

– $\alpha = \dfrac{\dot{V}_w}{\dot{V}_r}$ is the diafiltration ratio;

– J is the permeation flux density ($L.h^{-1} m^{-2}$);

– A_d is the membrane surface area required for diafiltration (m^2);

– VRR_d is the volume reduction factor at which diafiltration takes place.

We can write:

$$\dot{V}_p = A_d J \qquad [6.22]$$

and:

$$VRR_d = \frac{\dot{V}_f}{\dot{V}_r} = \alpha \frac{\dot{V}_f}{\dot{V}_w} \qquad [6.23]$$

where $\dot{V}_f$ is the volumetric flow rate of the system. By combining [6.21–6.23], we obtain:

$$A_d = \frac{\alpha}{VRR_d J} \dot{V}_f \qquad [6.24]$$

The membrane surface area A_d decreases as the product VRR_d J increases and therefore diafiltration is carried out where these latter is at a maximum (Figure 6.10).

Composition of fractions

The composition of concentrates obtained can be determined based on a mass balance:

– VRR is the volume reduction factor;

– X_f, X_r, X_p are the respective concentrations of a given constituent X ($g\ kg^{-1}$) in feed, retentate and permeate;

– $\dot{V}_f$, $\dot{V}_r$ and $\dot{V}_p$, and ρ_f, ρ_r and ρ_p are the volume flow rates ($m^3 s^{-1}$) and densities ($kg\ m^{-3}$) of the feed, the retentate and the permeate obtained.

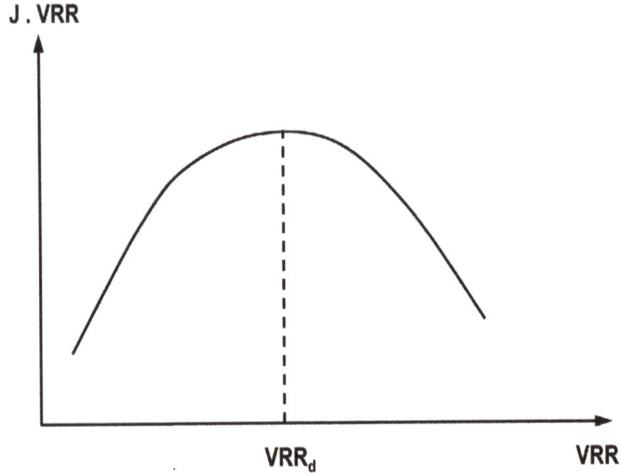

Figure 6.10. *Determination of the optimal volume reduction factor for diafiltration*

The mass balance for constituent X can be written as:

$$\rho_f \dot{V}_f X_1 = \rho_r \dot{V}_r X_r + \rho_p \dot{V}_p X_p \quad [6.25]$$

Knowing that $\dot{V}_p = \dot{V}_f - \dot{V}_r$, the following is obtained:

$$\rho_f \dot{V}_f X_f = \rho_r \dot{V}_r X_r + \rho_p \left[\dot{V}_f - \dot{V}_r\right] X_p$$

which leads to:

$$X_r = \frac{VRR}{\rho_r}\left[\rho_f X_f - \rho_p X_p\right] + \frac{\rho_p}{\rho_r} X_p \quad [6.26]$$

In a first approximation, it is possible to simplify this expression by assuming that the densities of the different liquids are the same. We therefore obtain:

$$X_r = VRR\left[X_f - X_p\right] + X_p \qquad [6.27]$$

It is essential to clearly define the limits of the system to which [6.26] can be applied. For example, in the case of diafiltration, this equation can only be applied after the addition of water, that is X_f is the concentration of element X after dilution.

6.2.2.2. *Size exclusion chromatography*

Size exclusion chromatography, also known as gel filtration, is a molecular separation technique based on differences in the transfer rate of molecules through a silica or polysaccharide gel (e.g. dextran). Dextran gels are in the form of porous beads and their porosity and pore characteristics depend on the degree of cross linking; these beads are hydrophilic and swell upon contact with water.

The principle of separation is shown in Figure 6.11; small molecules can penetrate inside the beads because their diameter is less than the pore size of the matrix while large molecules are excluded; large particles with a shorter path are eluted before small molecules. The elution volume varies with molecular weight (Figure 6.12).

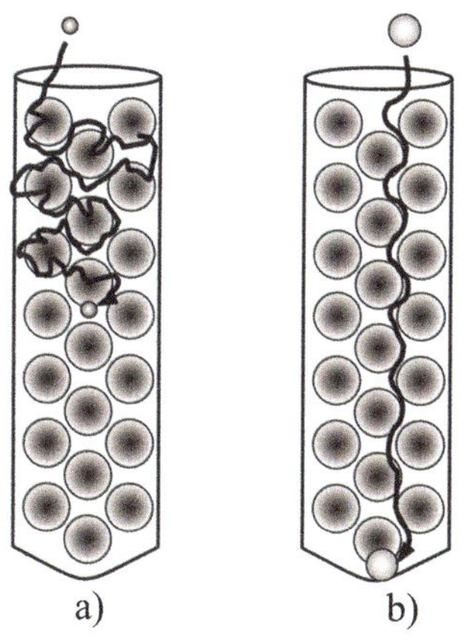

Figure 6.11. *Principle of size exclusion chromatography: path and transfer rate of small a) and large molecules b)*

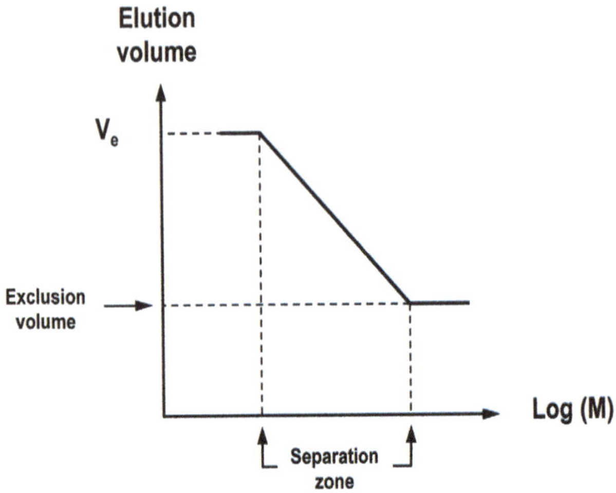

Figure 6.12. *Elution volume based on molar mass*

6.3. Separation by charge

6.3.1. *Physicochemical properties affecting molecular charge*

Many macromolecules have cationic and/or anionic groups making it possible for them to interact electrostatically. Charge type and density are highly dependent on the physicochemical conditions of the medium, in particular pH and ionic strength. As already mentioned, an acid group is mostly ionized at $pH > pK_a$, an amine group is mostly protonated and positively charged at $pH < pK_b$ and an amphoteric molecule (amino acid, peptide, protein) is usually positively charged at $pH < pI$ and negatively charged at $pH > pI$. In addition, ionic strength affects both pI and electrostatic bonds. Ionic strength and pH are two parameters that can increase or decrease the charge or even reverse it in the case of amphoteric molecules.

6.3.2. *Separation processes*

6.3.2.1. *Electrodialysis*

Electrodialysis is an electrochemical process that removes ionic species contained in a solution; this is achieved by transferring ions through semi-permeable membranes under the influence of an electric field. Alternating an ion exchange (M_A, positively charged) and cation exchange membranes (M_C, negatively charged) are used; one electrodialysis compartment consists of two membranes, an anion exchange and a cationic exchange membrane, which are perpendicular to the electric field (Figure 6.13). The C^+ cations in the ionic solution migrate towards the cathode: they cross the cation exchange membrane but are retained in the next compartment (concentrate stream) as they are repelled by the anion exchange membrane. Likewise, A^- anions migrating in the opposite direction towards the anode pass through anion exchange membranes and are repelled by cation exchange membranes. Anions and cations are extracted from the brine (concentrate stream). An electrodialysis device can have up to 1,000 dilute and concentrate compartments.

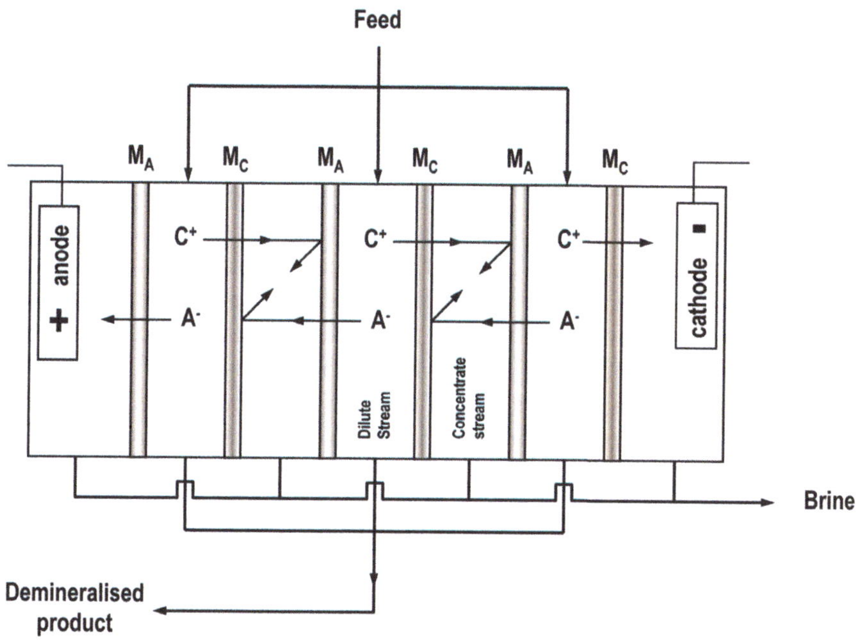

Figure 6.13. *Principle of electrodialysis*

Only ionic species are extracted but the rate of extraction depends on ionic mobility and molecular size: the extraction yield of mineral cations and anions is very high, but for organic molecules, it is highly size-dependent.

When solutions contain charged macromolecules, there is an accumulation of macromolecules on the surface of the charged membranes. In addition, demineralization causes a local reduction in the ionic strength of the solution (primary polarization in the boundary layer at the membrane), which can sometimes cause precipitation of inorganic salts or macromolecules on the membrane (secondary polarization) due to variations in electric current and pH. It is therefore highly recommended to limit the level of demineralization to avoid these limiting factors and prevent high ohmic resistance of the system, which increases energy costs.

This is a very suitable technique for partially demineralizing protein solutions to extract amino acids and organic acids (lactic acid, tartaric acid, malic acid, etc.). It is used industrially for the demineralization of whey and buttermilk, as well as grape juice and white wine (deacidification and removal of potassium tartrate).

6.3.2.2. *Ion exchange*

Ion exchange matrices consist of organic (polystyrene, cellulose, dextran) or inorganic matrices (silica), onto which positively or negatively charged functional groups are covalently grafted. They are insoluble in water and therefore easily recovered after contact with solutes. The most common functional groups used are SO_3^-, CO_2^-, $CH_2CO_2^-$ for cation exchangers and NH_3^+, $CH_2\text{-}NH\text{-}(C_2H_5)_2^+$, $N(CH_3)_3^+$ for anion exchangers; the type of exchanger used depends on the pH of the operation and the pK of the ionic species to be extracted. Separation is dependent on the properties of the matrices such as porosity, which depends on the level of crosslinking in the case of organic polymers, bead size, which can vary from 5 to 500 μm and binding capacity, which corresponds to the amount of ions that can bind to the exchange matrix (mEq g^{-1}). The level of porosity depends on molecular size, and bead size is determined based on a compromise: the smaller the particles, the faster the exchange and the greater the efficiency; however, small bead size generates higher costs and significant pressure drops in a fixed-bed exchanger.

The affinity of an exchanger for a charged molecule depends on:

– the pK of the functional group;

– the charge density of the ion to be extracted, which is based on the pK of the functional groups or the pI in the case of amphoteric molecules;

– the ion concentration.

The exchanger can be equilibrated to different pH levels:

– in an acidic or basic medium with the counter ion being H^+ in cation exchangers and OH^- in anion exchangers;

– in a neutral medium with K^+, Na^+ for cation exchangers and Cl^- for anion exchangers.

The exchange reaction in the case of a cation exchanger can be written as:

$$\begin{array}{cccc} R-B^+ & + A^+ \rightarrow & R-A^+ & + B^+ \\ [B_R] & [A_S] \leftarrow & [A_R] & [B_S] \end{array} \qquad [6.28]$$

where $[B_R]$, $[B_S]$ and $[A_R]$, $[A_S]$ are the concentrations of B^+ and A^+ cations bound to R exchanger and in solution, respectively. The selectivity constant K_B^A (dimensionless) is given by the equation:

$$K_B^A = \frac{[A_R][B_S]}{[B_R][A_S]} \qquad [6.29]$$

If $K_B^A > 1$,the exchanger has a preferential affinity for A, and the fixation of A^+ ions increases if the solution is regenerated with A_S; this is why a counter-current operation is preferred in the case of a semi-continuous or continuous process.

The ionic compound that is bound to the matrix can be eluted by changing the pH or ionic strength to reduce the charge density of the immobilized ion and/or decrease the energy of the electrostatic bonds. This is done by replacing the charged solute with another ion, either by applying increasing salt concentrations (commonly high molar NaCl solution) or using ions with a high affinity for the exchanger.

Ion exchangers can be operated in batch mode, continuous fed-batch mode possibly divided into a number of stages, fixed bed mode or a combination of batch and fixed bed modes, the first intended for fixation and the latter for elution. Fixed bed elution has the advantage of limiting the volume of eluent and obtaining a more concentrated elution of separate ionic species. After elution, the exchanger is regenerated.

Figure 6.14 shows the coupling of two continuously stirred reactors operating counter-currently with a fixed bed for rinsing, elution and regeneration.

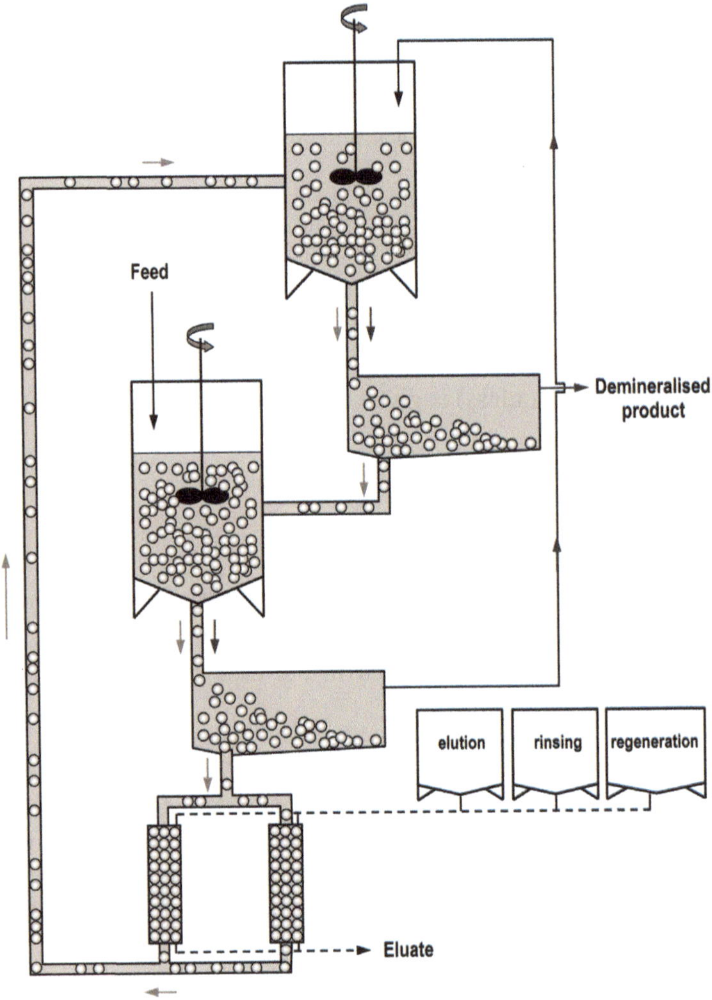

Figure 6.14. *Continuous stirred (two stage counter-current) reactors coupled with a fixed bed for elution*

Figure 6.15 describes a method for separating amphoteric molecules with different pI values; if the pH of the solution is between the two pI values, molecules with a pI below the pH of the solution are negatively charged and those with a pI above the pH of the solution are positively charged. By successively passing through an anion exchanger and a cation exchanger, charged molecules are separately bound to the matrices and neutral molecules pass through. Elution involves reversing the charge of the molecules by decreasing or increasing the pH through the addition of acid or sodium hydroxide, respectively.

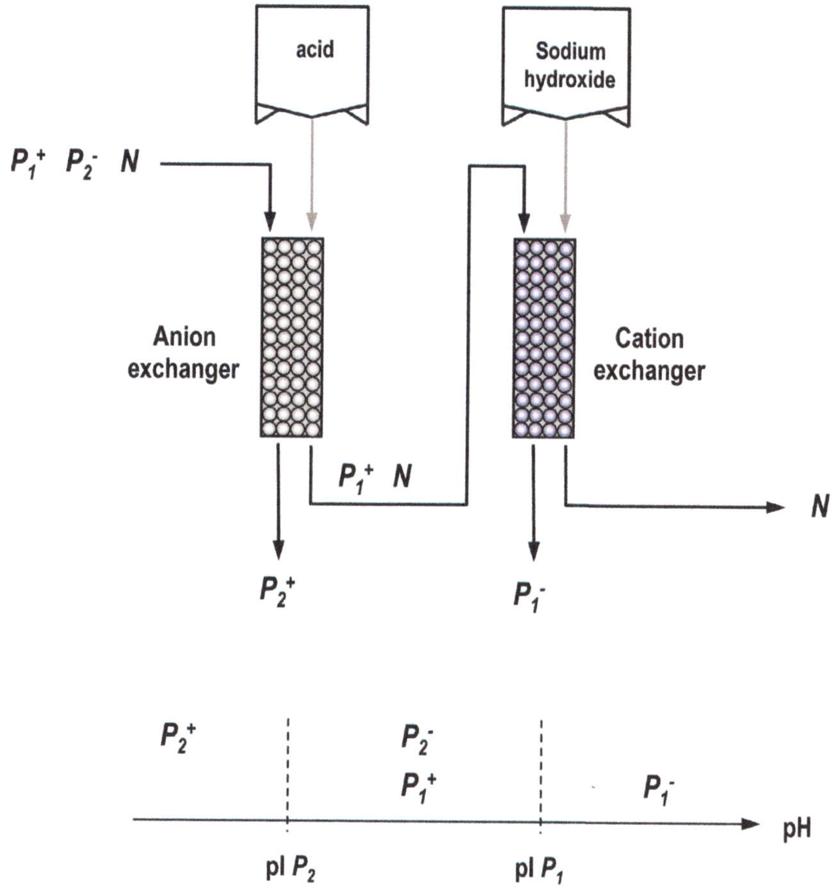

Figure 6.15. *Extraction of amphoteric molecules with different pI values using anion and cation exchange beds*

When attempting to separate two amphoteric molecules with similar pI values but with different affinities for the ion exchanger, both can be separated from each other in a fixed bed system; during feeding, the exchanger retains the two molecules and once all the sites are occupied, the molecule with the greater affinity replaces the one with less affinity. Under these conditions, the eluate only contains the molecule with less affinity, and when the exchanger reaches saturation, it is then rinsed and eluted. This separation by displacement has the advantage of limiting the use of eluents (frontal chromatography).

6.4. Separation by affinity chromatography

6.4.1. *Immobilization of ligands*

Some molecules can specifically bind to a ligand; these ligands may be antibodies, enzyme substrates, glycosides or minerals (transition minerals). However, it is necessary to bind the ligand to an insoluble matrix without altering its affinity properties to the target molecule(s). The most common matrices are resins that are inert to the constituents of the medium. When the ligand or the compound to be isolated are macromolecules whose spatial structures determine their affinity properties, immobilization on the matrix can generate steric hindrance, which may prevent the formation of the complex.

Resins are commercially available, which have chemical groups that are easily activated, possibly positioned at the end of a "spacer arm" of varying length to reduce steric hindrance and facilitate access to the ligand. In most cases, the ligand is covalently linked to amino or carboxyl groups (Figure 6.16).

$$\boldsymbol{R}\text{-O-CH2} - \text{CO} - \text{NH} - (\text{CH2})\ \text{n} - \text{NH2}$$

$$\boldsymbol{R}\text{-O-CH2} - \text{CO} - \text{NH} - (\text{CH2})\ \text{n} - \text{NH} - \text{CO} - \text{CH2} - \text{CH2} - \text{COOH}$$

Figure 6.16. *Types of ligands used in affinity chromatography*

6.4.2. *Separation*

Separation by affinity chromatography from a complex medium occurs in three stages: binding, washing and elution (Figure 6.17). Immobilization yield and losses during washing depend on the affinity constant K of the molecule M to the ligand L, which may depend on the physicochemical conditions (pH and ionic strength) of the medium:

$$M + L \leftrightarrow ML$$

$$K = \frac{[ML]}{[M]\ [L]} \qquad [6.30]$$

During the operation, it is important to control the physicochemical conditions of the washing solution. Elution can be achieved either by modifying the physicochemical conditions of the eluent (pH and ionic strength) or by adding a free ligand to the eluent, which competes with the immobilized ligand. Strong reactions between the molecule to be extracted and the resin can result in harsh elution conditions and, in extreme cases, can denature the target molecule.

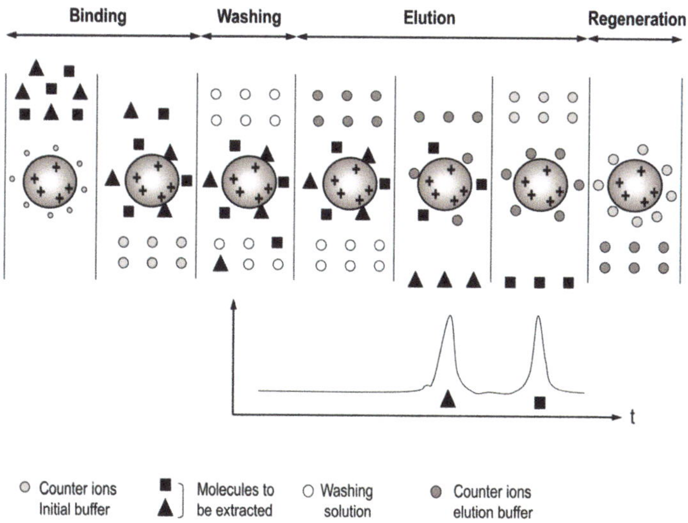

Figure 6.17. *Separation by affinity chromatography from a complex medium*

As already outlined, three types of methods are used in ion exchange processing: stirred batch, fixed bed and a combination of both batch and fixed bed. Stirred batch involves exposing the resin and the mixture to be treated until complex formation; the resin is then recovered by decantation or filtration and washed. The immobilized molecule is desorbed under conditions defined above either in batch or fixed bed.

Desorption in a fixed bed has the advantage of reducing the volume of eluent used and obtaining a more concentrated eluate. The fixed bed system, which involves filling the resin into a column and adding the mixture to be separated, is better suited when volumes are lower and resins are brittle.

Given the cost of immobilized ligands, this technique is only used for the extraction of high-value added components present in small quantities; the advantage of this method is that it generates a very similar by-product to the raw material from which the components are extracted.

6.5. Extraction of lipophilic molecules

6.5.1. *Molecular partition between two immiscible phases*

The division of a solute A into two immiscible liquid phases, one aqueous and the other organic, is a defined equilibrium by the partition coefficient K (dimensionless):

$$A_{aq} \rightleftarrows A_{org}$$

$$K = \frac{a_{A_{org}}}{a_{A_{aq}}} \qquad [6.31]$$

where a_{Aorg} and a_{Aaq} are the activities of solute A in the organic and aqueous phases respectively. In the case of diluted solutions and uncharged molecules, activity and concentration are equal and K can be expressed as a function of the concentration (C_{Aorg} and C_{Aaq}) or mass (m_{Aorg} and m_{Aaq}) of solute A in each phase:

$$K = \frac{C_{A_{org}}}{C_{A_{aq}}} = \frac{m_{A_{org}}}{V_{org}} \frac{V_{aq}}{m_{A_{aq}}} \quad [6.32]$$

where V_{org} and V_{aq} are the volumes of the organic and aqueous phases respectively. Given the mass conservation of the solute:

$$m_{A_{org}} + m_{A_{aq}} = m_0 \quad [6.33]$$

The residual mass of the solute in the aqueous phase is therefore:

$$m_{A_{aq}} = \frac{m_0 V_{aq}}{K V_{org} + V_{aq}} \quad [6.34]$$

More of the solute can be extracted if the solvent volume is divided into several successive extractions. For example, for:

$K = 10$

$V_{org} = 0.2V_{aq}$

The residual mass of the solute is $m_{A_{aq}} = \frac{m_0}{3}$.

If two successive extractions are carried out with a volume of solvent:

$V_{org} = 0.1\ V_{aq}$

The residual mass in the first extraction is $m_{A_{aq}1} = \frac{m_0}{2}$, and in the second extraction $m_{A_{aq}2} = \frac{m_0}{4}$.

Figure 6.18 shows a solvent extraction with solvent recovery distillation.

Extraction can be a multi-stage, counter-current process (Figure 6.19), which produces a better extraction yield.

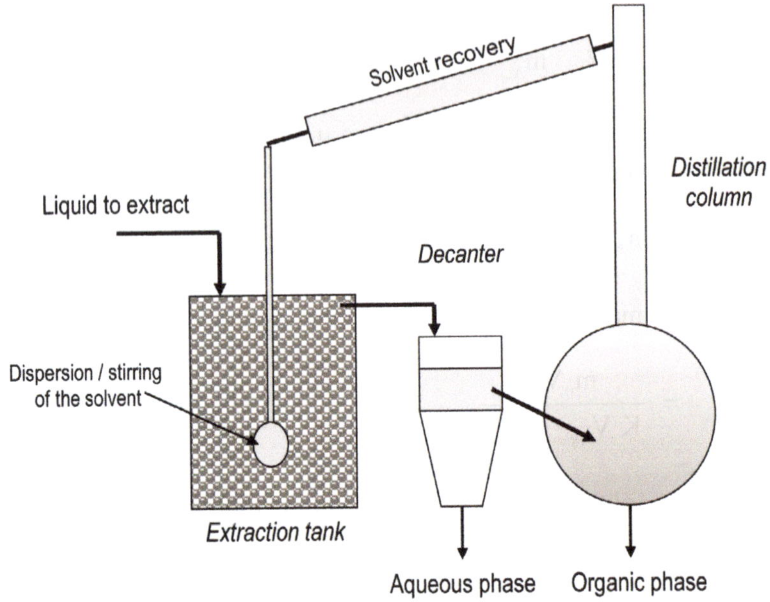

Figure 6.18. *Solvent extraction with solvent recovery*

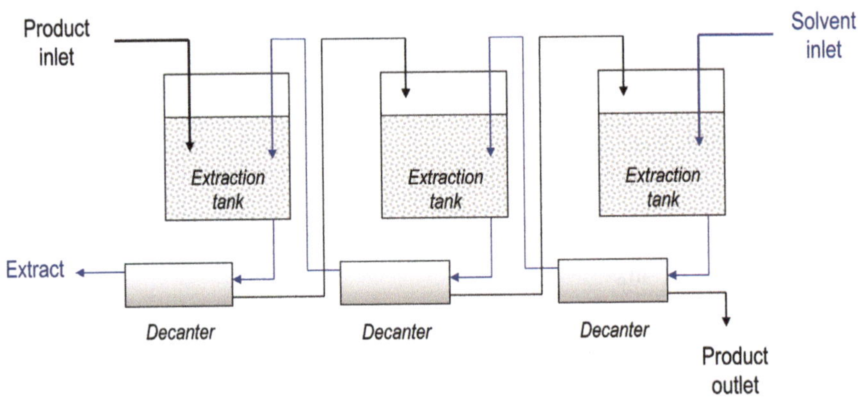

Figure 6.19. *Counter-current liquid-liquid extraction*

Organic solvents may pose health and safety risks and/or exhibit toxicity. Supercritical CO_2 is a solvent that has many advantages, but its use is currently limited because of its high cost. Due to its density, it has a solvent power close to liquid CO_2 and high diffusivity properties; it therefore has a penetration power close to that of gaseous CO_2. CO_2 removal is achieved by expansion and recovery in the supercritical state by recompression (Figure 6.20).

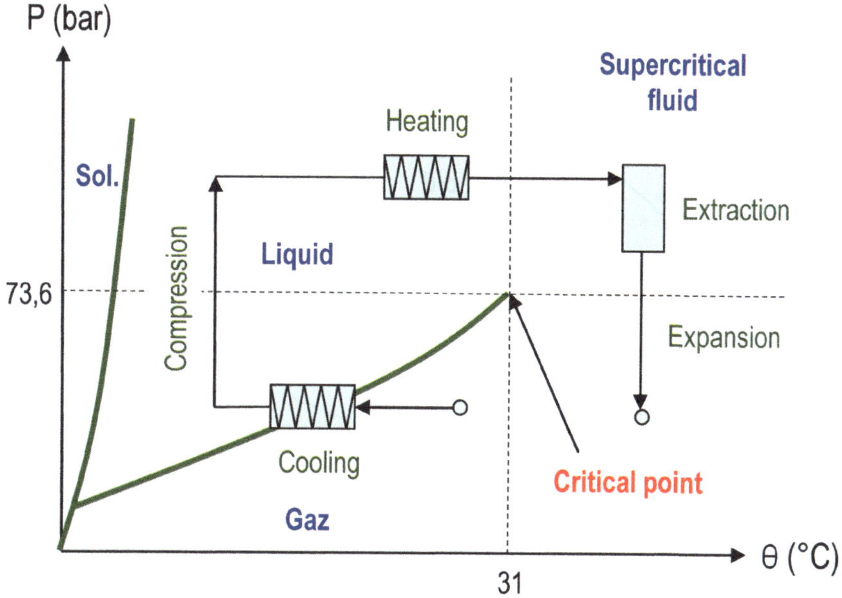

Figure 6.20. *Supercritical CO_2 extraction*

6.6. Biotransformation and its use in separation

When it is not feasible to separate two molecules with very similar physicochemical properties using the methods described in the previous section, the compounds can be subjected to biological transformation. We have already mentioned the enzymatic hydrolysis of proteins or polysaccharides when purifying one of the hydrolyzed fraction from these components either by ultrafiltration or size exclusion chromatography, for example the separation of two carbohydrates (G_1 and G_2) and two enantiomers from a racemate.

In the case of carbohydrates (Figure 6.21), some microorganisms and enzymes can specifically transform the carbohydrate to be removed; the biomass can be isolated by centrifugation or microfiltration, acids by ion exchange or electrodialysis, and alcohols by solvent extraction or distillation.

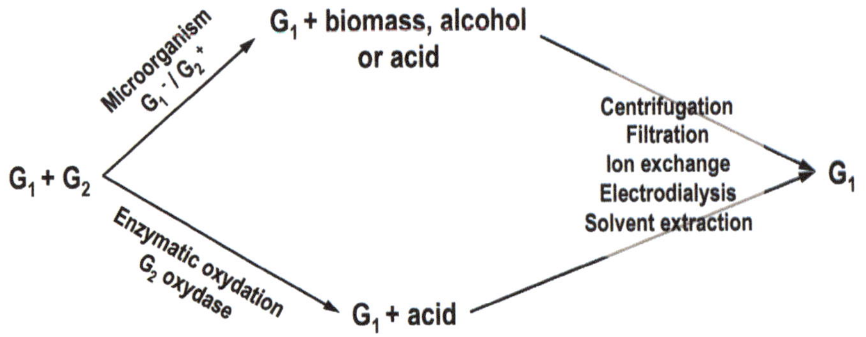

Figure 6.21. *Separation of carbohydrates by biotransformation*

The separation of two enantiomers (R and S; Figure 6.22) may be carried out after chemical derivatization of the racemate followed by stereospecific enzymatic transformation. If, for example, the molecules contain an acid or an alcohol function, chemical esterification with a molecule B followed by stereospecific enzymatic hydrolysis is carried out. The reaction mixture is then composed of ester, alcohol and acid, which is easier to separate.

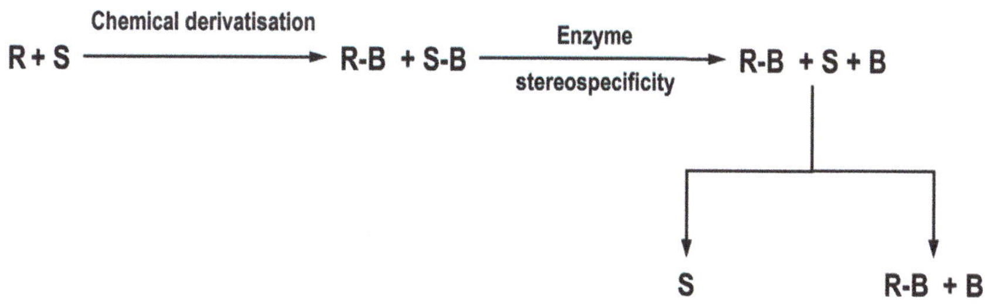

Figure 6.22. *Separation of two enantiomers by biotransformation*

7

Biotransformation and Physicochemical Processing

The development of the primary processing industry is attributable to a better understanding of the biochemical and physicochemical characteristics of agricultural raw materials and food industry by-products as well as technical advances made in the field of separation processes. The initial main activity of this industry was the separation of different components in order to better exploit their nutritional and/or techno-functional properties. It has gradually sought to develop functional ingredients to meet the demands of the secondary processing industry. Based on the demand for functionality, it uses all biological and physicochemical means to transform food components from separation processes and generate new functionalities.

7.1. Biotransformation

7.1.1. *Biological agents*

The biotransformation of food components can be achieved using microorganisms or enzymes. Whole cells are used in the biotransformation process if several enzymatic reactions are involved or if certain co-factors are required that are expensive or difficult to generate; this is the case with the production of alcohol, organic acids, amino acids or vitamins. However, when biotransformation involves the action of an enzyme that is easy to produce, this approach is preferred. Individual enzymes are generally easier

---

Chapter written by Romain JEANTET.

to use than cells, which require greater energy input and yield products that are more complex to purify.

A large number of food-grade enzymes currently exist in the food processing industry: these enzymes can be of animal (chymosin, pepsin, trypsin, chymotrypsin, lysozyme, etc.), plant (papain, ficin, bromelain, amylase, lipoxygenase, etc.) or microbial origin. The abundance of microbial enzymes on the market is due to the greater regularity of supply, low production costs and increasing opportunities in genetically improved strains. Several microorganisms are currently permitted in the production of enzymes; the most common microbial genera include *Bacillus*, *Aspergillus*, *Saccharomyces* and *Kluyveromyces*. The most widely used enzymes are hydrolases (protease, lipase, amylase, lactase, invertase, pectinase, etc.) isomerases (glucose isomerase) and oxidases (glucose oxidase, lipoxygenase).

Microbial enzymes are obtained by fermentation; extracellular enzymes are extracted directly from the culture medium while endocellular enzymes require prior cell lysis. As regards plant and animal preparations, tissues are subjected to physical treatment (grinding, pressing, sonication, freeze-thaw cycle); the release of cell contents can also be promoted by chemical lysis (pH, ionic strength) or enzymatic lysis (lysozyme, pectinase, cellulose). The purification of enzymes involves fractionation techniques based on the differences in properties between the enzyme and the other components in the medium as already described in the previous chapter (solubility, molar mass, electrical charge). Some analytical methods are used to assess the purity (electrophoresis, chromatography) and activity of the extract (spectrophotometry under standard conditions); the preparations are often dried by lyophilization to preserve their activity. The cost of the preparations is largely dependent on the level of purity achieved: the presence of enzymatic activity other than that desired can be problematic in certain applications.

7.1.2. *Kinetics of biotransformation*

7.1.2.1. *Microbial kinetics*

Microbial growth can be divided into five phases as shown in Figure 7.1:

– lag phase (❶): metabolic adaptation of microorganisms to the conditions of the medium;

– exponential growth phase (❷): suitable conditions for growth – no limiting substrates – no inhibition by metabolites;

– deceleration phase (❸): unsuitable conditions for growth – substrates and/or factors limiting growth– inhibition by metabolites;

– stationary phase (❹): halt in growth;

– decline phase (❺): cell death.

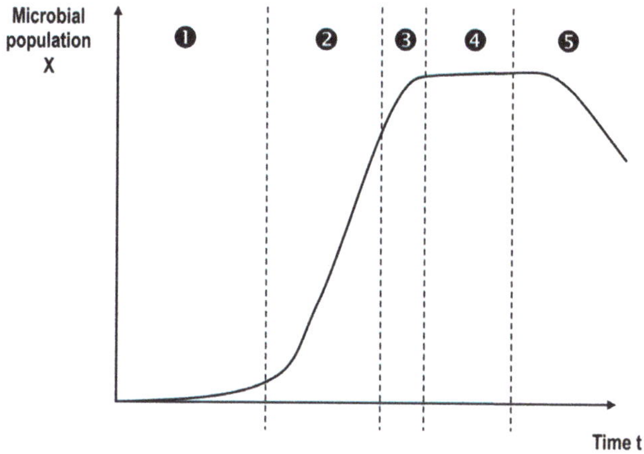

Figure 7.1. *Microbial growth*

The microbial growth rate v_x (g of biomass.h^{-1}) is proportional to cell concentration X (first order reaction):

$$v_x = \frac{dX}{dt} = \mu X \qquad [7.1]$$

μ is the specific growth rate (h^{-1}). Integrating this equation between time t_1 (cell concentration X_1) and t_2 (cell concentration X_2) gives:

$$\ln\left(\frac{X_2}{X_1}\right) = \mu\ (t_2 - t_1) \qquad [7.2]$$

$$X_2 = X_1\ e^{\mu\,(t_2 - t_1)} \qquad [7.3]$$

The time required to double the microbial population is known as the generation time (t_G; h):

$$t_G = \frac{\ln 2}{\mu} = \frac{0.69}{\mu} \qquad [7.4]$$

The generation time is about 0.3 h for bacteria, 1.5 h for yeast and 3 h for fungi. The rate constant μ decreases with consumption of the substrate (Figure 7.2).

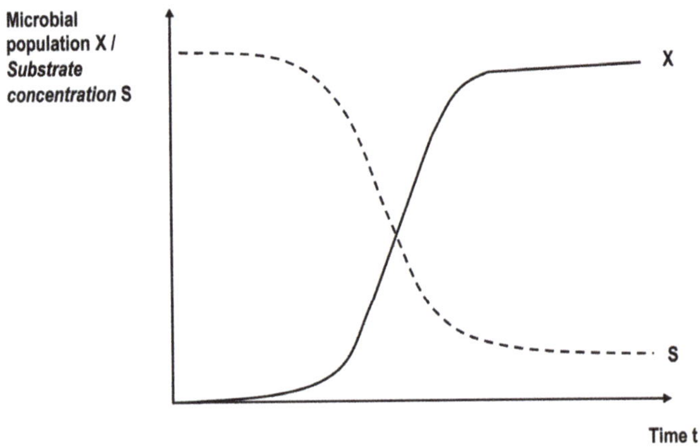

Figure 7.2. *Change in biomass X as a function of the concentration of substrate S*

The Monod model defines the effect of the substrate concentration (S) on the specific growth rate (Figure 7.3):

$$\mu = \frac{\mu_{max} \; S}{K_S + S} \qquad [7.5]$$

where μ_{max} is the asymptotic value of the specific growth rate μ, and K_S is the threshold or Monod constant corresponding to the substrate concentration for which $\mu = \frac{1}{2}\mu_{max}$. K_S is expressed based on the concentration unit (mol.m^{-3} or kg m^{-3}). By combining [7.1] and [7.5], we get the biomass production rate:

$$v_x = \frac{\mu_{max} \; S}{K_S + S} \; X \qquad [7.6]$$

Thus:

– When S >> K_S:

$$v_x = \mu_{max} \; X \qquad [7.7]$$

This relationship consequently corresponds to a zero order reaction with respect to the substrate and a first order reaction with respect to the biomass;

– When S << K_S:

$$v_x = \frac{\mu_{max}}{K_S} \; S \; X \qquad [7.8]$$

The reaction follows first order kinetics for both the substrate and the biomass.

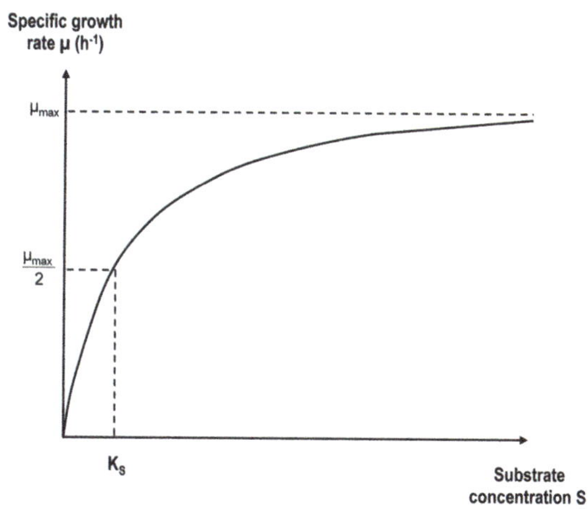

Figure 7.3. *Effect of the substrate concentration on the specific growth rate*

Knowing that the substrate concentration (S) decreases as biomass X is formed, v_x therefore reaches a maximum. At the start of growth (S >> K_S), the rate increases with biomass; however when S decreases and tends towards K_S, the specific rate drops faster than the biomass increases and v_x decreases.

Biomass X can be expressed as a function of the amount of substrate consumed ($S_0 - S$) for yield Ψ (g of biomass per g of substrate):

$$X = X_0 + \psi \ (S_0 - S) \quad [7.9]$$

X_0 and S_0 are the initial concentrations of biomass and substrate. By combining [7.6] and [7.9], v_x can be expressed as:

$$v_x = \frac{\mu_{max} \ S}{K_S + S} \ [X_0 + \psi \ (S_0 - S)] \quad [7.10]$$

To find the maximum rate, we differentiate v_x with respect to S:

$$v'_x = \frac{dv_x}{dS} = -\frac{\mu_{max} \ S}{K_S + S} \psi + \frac{\mu_{max} \ (K_S + S) - \mu_{max} \ S}{(K_S + S)^2} \ [X_0 + \psi \ (S_0 - S)] \quad [7.11]$$

Leading to:

$$v'_x = \mu_{max} \ \psi \ \frac{- S^2 - 2 K_S \ S + K_S \ \dfrac{X_0}{\psi} + K_S \ S_0}{(K_S + S)^2} \quad [7.12]$$

This derivative is zero for:

$$S = - K_S \pm \sqrt{K_S^2 + K_S \ (\frac{X_0}{\psi} + S_0)} \quad [7.13]$$

Where $\dfrac{X_0}{\psi} << S_0$, conditions are optimal for:

$$\frac{S}{K_S} = \sqrt{1 + \frac{S_0}{K_S}} - 1 \quad [7.14]$$

The maximum rate is therefore dependent on the initial concentration of biomass and substrate as shown in Figure 7.4.

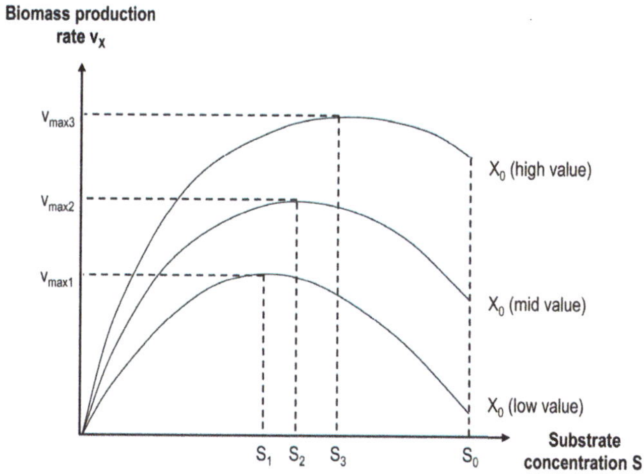

Figure 7.4. *Substrate concentration for a maximum rate based on the initial biomass concentration*

The kinetic parameters can be determined from the Monod equation, plotted as a straight line (Figure 7.5):

$$\frac{X}{v_x} = \frac{K_S}{\mu_{max} S} + \frac{1}{\mu_{max}} \qquad [7.15]$$

Where two substrates are limiting, the model is written as:

$$\mu = \mu_{max} \left(\frac{S_1}{K_S + S_1}\right) \left(\frac{S_2}{K_S + S_2}\right) \qquad [7.16]$$

When a metabolite P is inhibiting:

$$\mu = \mu_{max} \left(\frac{S}{K_S + S}\right) \left(\frac{1}{1 + \dfrac{P}{K_P}}\right) \qquad [7.17]$$

P is the metabolite concentration, and K_p is its inhibition constant (which is expressed by the same unit as concentration P). Many other growth models have been proposed either with or without inhibitors.

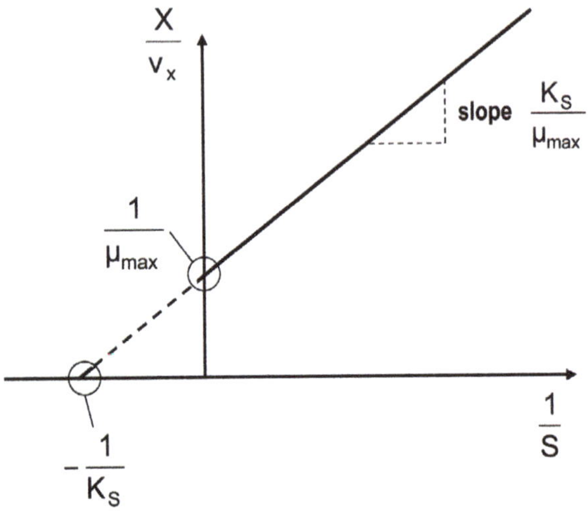

Figure 7.5. *Monod equation in the form of a straight line*

7.1.2.2. *Enzymatic kinetics*

Enzymatic kinetics is often represented using the Michaelis–Menten model, which takes into account the rate of formation and disappearance of the enzyme (E) – substrate (S) complex according to the following equilibria:

$$E + S \underset{K_2}{\overset{K_1}{\rightleftarrows}} ES \xrightarrow{K_3} P$$

The formation rate of the ES complex is:

$$\frac{d\,ES}{dt} = k_1 \; ES \qquad [7.18]$$

where k_1, a second order rate constant, is expressed according to the concentration ($kg^{-1}\ m^3\ s^{-1}$ or $mol^{-1}\ m^3\ s^{-1}$). The rate of disappearance of this complex is:

$$-\frac{d\,ES}{dt} = (k_2 + k_3)\,ES$$

where k_2 and k_3 are first order rate constants and are expressed in s^{-1}. In the steady state:

$$k_1\ E\ S - (k_2 + k_3)\,ES = 0 \qquad [7.19]$$

If E_0 and E are respectively the initial and equilibrium concentrations of the enzyme, one can write:

$$E_0 = E + ES$$

By replacing E by $[E_0 - ES]$ in [7.19], we obtain the concentration of the ES complex:

$$ES = \frac{E_0\ S}{\frac{k_2 + k_3}{k_1} + S} \qquad [7.20]$$

The ratio of the first order rate constants ($k_2 + k_3$) and the second order constant (k_1) is known as the Michaelis–Menten constant K_m, which has the dimension of a concentration ($mol\ m^{-3}$ or $kg\ m^{-3}$):

$$\begin{cases} ES = \dfrac{E_0\ S}{K_m + S} \\ K_m = \dfrac{k_2 + k_3}{k_1} \end{cases} \qquad [7.21]$$

The enzymatic reaction rate v ($mol\ m^{-3}\ s^{-1}$ or $kg\ m^{-3}\ s^{-1}$) can be expressed by the Michaelis–Menten equation:

$$v = \frac{dP}{dt} = k_3\ ES = k_3\ \frac{E_0\ S}{K_m + S} \qquad [7.22]$$

The rate is maximum when all of the enzyme is in complex form, that is $E_0 = ES$:

$$v_{max} = k_3 \; E_0 \qquad [7.23]$$

By combining [7.22] and [7.23], we obtain:

$$v = \frac{v_{max} \; S}{K_m + S} \qquad [7.24]$$

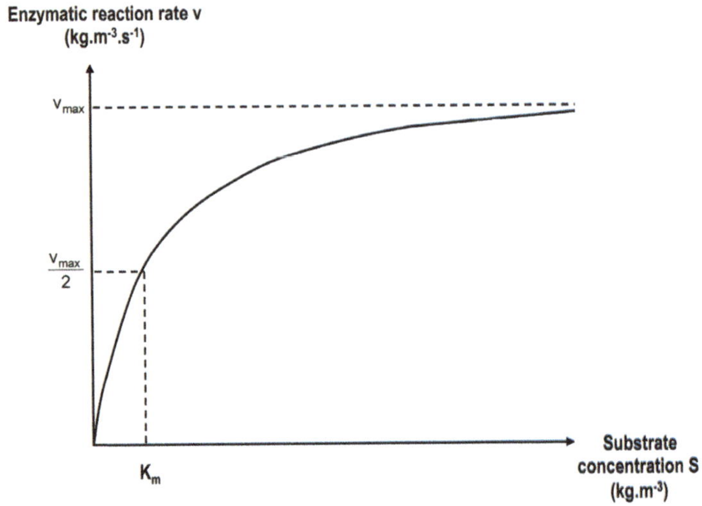

Figure 7.6. *Effect of substrate concentration on the reaction rate*

The Michaelis–Menten constant K_m is equivalent to the substrate concentration at which the rate $v = \frac{V_{max}}{2}$ (Figure 7.6); it is therefore close to the Monod constant with regard to the specific growth rate, as previously mentioned (equation [7.5]):

– If $S >> K_m$:

$$v = k_3 \; E_0 \qquad [7.25]$$

This reaction therefore follows first order kinetics;

– If $S << K_m$:

$$v = k_3 \frac{E_0 \ S}{K_m} \qquad [7.26]$$

This reaction therefore follows second order kinetics.

Kinetic parameters (K_m and v_{max}) are determined from the initial rate values at different substrate concentrations; plotting a linear representation of the Michaelis–Menten model is obtained by $\frac{1}{v}$ as a function of $\frac{1}{S}$ (Lineweaver–Burk plot):

$$\frac{1}{v} = \frac{K_m + S}{v_{max} \ S} = \frac{K_m}{v_{max}} \frac{1}{S} + \frac{1}{v_{max}} \qquad [7.27]$$

This corresponds to a straight line with a slope of $\frac{K_m}{v_{max}}$ and an intercept of $\frac{1}{v_{max}}$ (Figure 7.7).

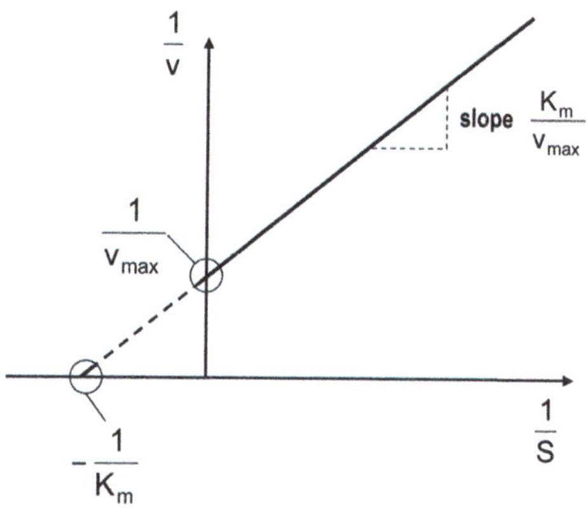

Figure 7.7. *Lineweaver-Burk plot of the Michaelis–Menten equation*

In the presence of an inhibitor (concentration I) capable of binding to the enzyme or the enzyme/substrate complex (association constant K_i), different types of competition can be encountered (Figure 7.8):

– competitive: inhibitor (I) forms a complex with the enzyme (EI);

– uncompetitive: inhibitor binds to the enzyme-substrate complex (ES) to give a new complex (ESI);

– non-competitive: the inhibitor forms complexes with both the enzyme and the complex (ES) to give complexes (EI) and (ESI).

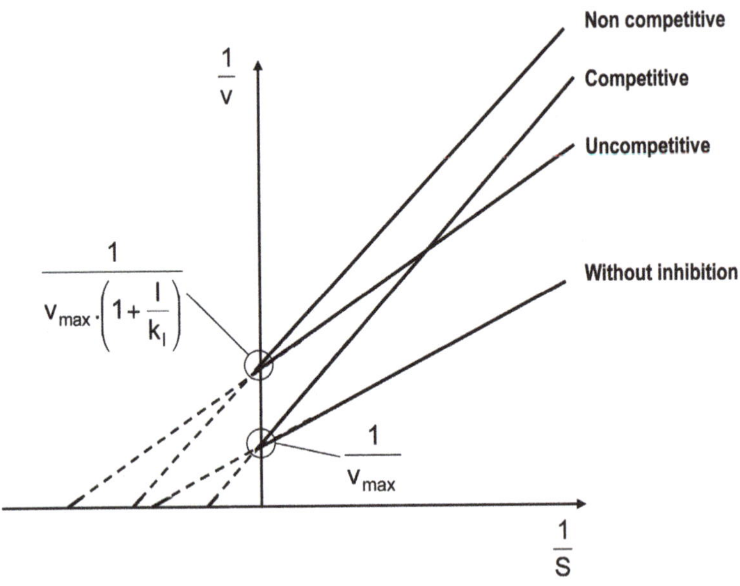

Figure 7.8. *Lineweaver–Burk plot in the presence of inhibitors*

In some cases, inhibition occurs when excessive amounts of substrate are present; the excess substrate forms a complex with the initial complex (ES) to give the complex (ESS).

7.1.2.3. *Kinetics in heterogeneous systems*

Enzymes and microorganisms can be used in stirred reactors. However, some reactors use immobilized microorganisms and enzymes on solid supports or encapsulated in particulate matrices. Immobilization on a solid support can be achieved by either covalent binding or non-covalent adsorption of enzymes to polymers (polyacrylamide gels, alginate beads, etc.).

These immobilized or encapsulated enzymes and microorganisms are used in stirred or fixed bed reactors. It is difficult to determine the optimal conditions for biochemical reactions in heterogeneous media because the physicochemical characteristics in the microenvironment of the biological reactants may be different to those in the bulk media.

The matrices used to immobilize or encapsulate the biocatalysts are usually cationic or anionic polymers. Their charge can generate ion partition effects (Figure 7.9), which create pH and ionic (substrate or reaction product) gradients: thus, the solute concentrations and pH of the reaction environment are different to those of the medium.

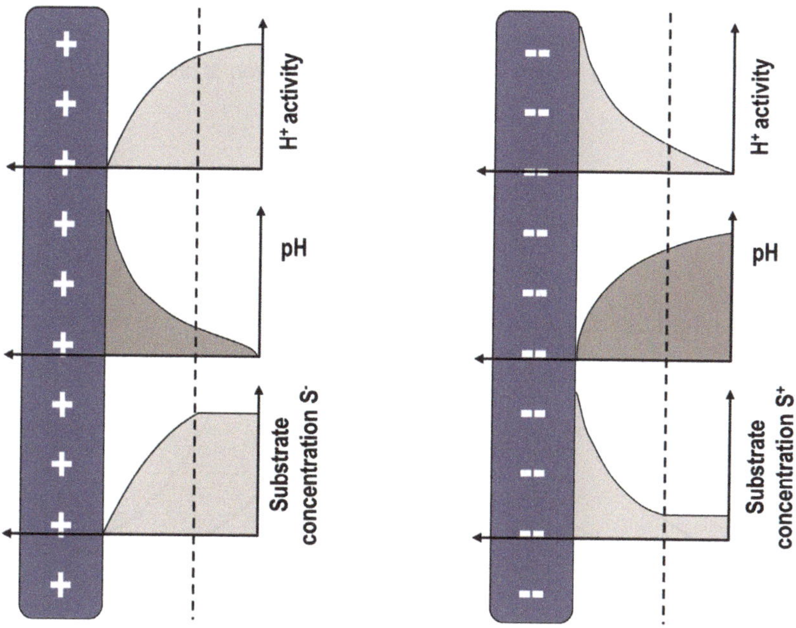

Figure 7.9. *Ion partition effect: gradients of H^+ activity, pH and concentration of negatively-charged substrate S at the interface of a charged matrix*

When using a positively-charged cationic matrix, the optimal pH of the reaction in heterogeneous media will be less than that in homogeneous media, and vice versa in the case of a polyanionic matrix as shown in Figure 7.10. These ion partition effects can be reduced by increasing the ionic strength of the medium as shown in Figure 7.11.

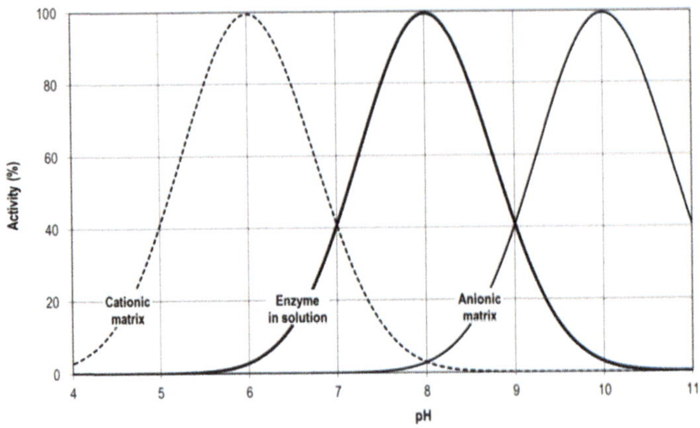

Figure 7.10. *Trypsin activity as a function of pH*

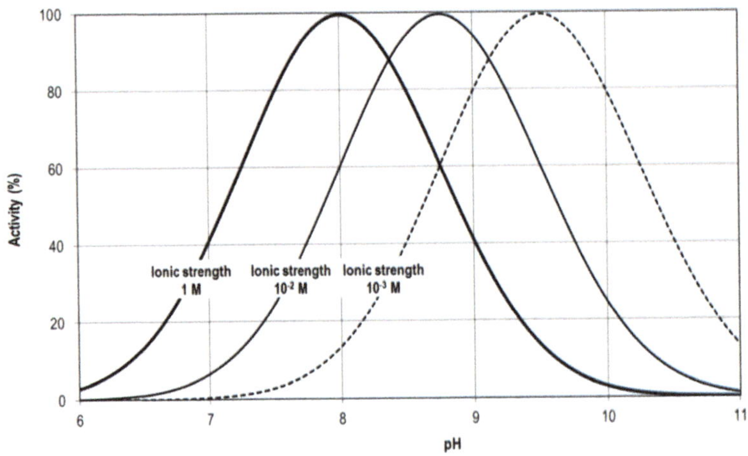

Figure 7.11. *Effect of ionic strength on the activity of trypsin immobilized on an anionic matrix*

Another phenomenon that may alter the kinetic parameters in heterogeneous media is the diffusion limit of substrates or reaction products in matrices or at the interfaces of solid and liquid phases. The consumption or conversion of the substrate in the microenvironment of biological agents generates a concentration gradient and the result is a transfer from the medium to the microenvironment: if the rate of transfer is greater than the rate of biological reactions, the substrate concentration in the microenvironment and the medium will be similar; otherwise it may create a concentration gradient, for example the level of substrate near the enzyme or microorganism can be very low compared to that of the medium. Given an interface layer thickness l and a surface A, in which a concentration gradient $\frac{\Delta S}{l}$ occurs in the steady state, the mass transfer is given by Fick's law (see Chapter 1):

$$\dot{m} = \frac{dm}{dt} = -A\ D_m\ \frac{\Delta S}{l} \qquad [7.28]$$

D_m is the diffusion coefficient of the material and is expressed as $m^2\ s^{-1}$ when the difference in substrate concentration ΔS between the medium (S_M) and the microenvironment (S_μ) is expressed in kg m^{-3}. The amount of transferred material dm results in a change in substrate concentration dS. Therefore:

$$dm = V\ dS = A\ l\ dS \qquad [7.29]$$

V is the interface volume. By combining [7.28] and [7.29], we obtain the rate of change of the substrate concentration:

$$\frac{dS}{dt} = -\frac{D_m}{l^2}\ \Delta S \qquad [7.30]$$

In the case of an enzymatic reaction, this can be combined with the Michaelis–Menten equation [7.24]:

$$-\frac{dS}{dt} = \frac{v_{max}\ S_\mu}{K_m + S_\mu} = \frac{D_m}{l^2}\ (S_M - S_\mu) \qquad [7.31]$$

This equation can be dimensionless where:

$$\begin{cases} \sigma_\mu = \dfrac{S_\mu}{K_m} \\ \sigma_M = \dfrac{S_M}{K_m} \\ \phi = \dfrac{V_{max}}{K_m} \dfrac{l^2}{D_m} \end{cases} \quad [7.32]$$

[7.31] becomes:

$$\sigma_M - \sigma_\mu = \frac{\phi \ \sigma_\mu}{1 + \sigma_\mu} \quad [7.33]$$

Depending on the case, [7.33] can be simplified:

– if $S_M << K_m$, we obtain σ_M and $\sigma_\mu << 1$:

– if $\varphi << 1$, $\sigma_\mu = \sigma_M$ [7.34]

– if $\varphi >> 1$, $\sigma_\mu = \dfrac{\sigma_M}{\phi}$ [7.35]

– if $S_M >> K_m$:

$$\sigma_\mu^2 + (\phi + 1 - \sigma_M) \ \sigma_\mu - \sigma_M = 0 \quad [7.36]$$

Knowing S_M, K_m and φ, it is possible to calculate S_μ.

To limit the difference in concentration between the medium and the reaction environment, φ must be reduced either by lowering the enzyme concentration or increasing the diffusion transfer (agitation to reduce the interface layer thickness *l*, which is a transfer limiting factor).

7.1.3. *Bioreactors*

The mode of operation in an industrial-scale bioreactor may be batch (working in cycles) or continuous, either in stirred tanks or in fixed-beds. There are other intermediate modes such as fed-batch, which can improve efficiency compared to the conventional batch.

7.1.3.1. *Batch reactor*

The use of batch reactors in enzymatic conversion is relatively simple and does not require expensive equipment. It usually consists of a tank with an agitator and thermostat to control reaction temperatures. Some reactions require a pH control system to maintain the optimal pH range or compensate for reaction-induced pH changes (e.g. protein hydrolysis). These reactors can operate with free enzymes in solution or immobilized enzymes held in suspension by agitation. The advantage of immobilized enzymes is that the enzyme preparation can easily be recovered for reuse by sedimentation or filtration. Cross-flow filtration (mainly ultrafiltration) is also commonly used to recover biological agents in the case of enzymes in solution provided the reaction products are of low molecular weight.

The operation of a batch enzyme reactor can be modeled on the Michaelis–Menten equation [7.22]:

$$-\frac{dS}{dt} = \frac{v_{max}\ S}{K_m + S} = \frac{k\ E_t\ S}{K_m + S} \qquad [7.37]$$

where E_t is the total enzyme concentration (free and complexed to the substrate) and k is the first order rate constant for the appearance of products (s^{-1}). Therefore:

$$-(K_m + S)\ \frac{dS}{S} = k\ E_t\ dt \qquad [7.38]$$

Integrated from t = 0 (S_0 concentration) to t (S concentration):

$$K_m\ \ln\left(\frac{S_0}{S}\right) + (S_0 - S) = k\ E_t\ t \qquad [7.39]$$

In addition, the conversion rate χ of the substrate (dimensionless) is defined by:

$$\chi = \frac{S_0 - S}{S_0} \qquad [7.40]$$

[7.39], expressed as a function of χ, becomes:

$$S_0 \ \chi - K_m \ \ln(1 - \chi) = k \ E_t \ t \qquad [7.41]$$

The kinetic parameters of the reaction can therefore be plotted according to Figure 7.12.

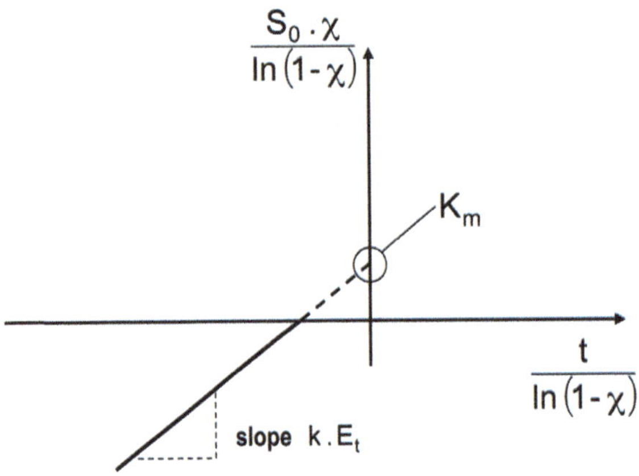

Figure 7.12. *Determination of kinetic parameters in a batch reactor*

Microbial batch reactors can also be modeled on the growth rate of microorganisms; in the absence of any inhibitors, the model can be expressed according to the Monod equation [7.6].

$$\frac{dX}{dt} = \frac{\mu_{max} \ S}{K_S + S} \ X \qquad [7.42]$$

Leading to:

$$dt = \frac{K_S}{\mu_{max}} \frac{dX}{S\,X} + \frac{1}{\mu_{max}} \frac{dX}{X} \quad [7.43]$$

If Ψ is the efficiency of converting the substrate, S_0 is the initial concentration and S is the concentration at time t, the microbial population X and its variation dX are:

$$X = X_0 + \psi\,(S_0 - S) \quad [7.44]$$

$$dX = -\psi\,dS \quad [7.45]$$

By combining [7.43–7.45], and integrating from S_0 to S, we obtain:

$$t = \frac{1}{\mu_{max}} \frac{K_S}{\frac{X_0}{\psi} + S_0} \left[\ln\left(\frac{S_0}{S}\right) + \ln\left(\frac{X}{X_0}\right) \right] + \frac{1}{\mu_{max}} \ln\left(\frac{X}{X_0}\right) \quad [7.46]$$

When $M = \dfrac{K_S}{\frac{X_0}{\psi} + S_0}$, [7.46] becomes:

$$t = \frac{1}{\mu_{max}} \left[M \ln\left(\frac{S_0}{S}\right) + (M+1) \ln\left(\frac{X}{X_0}\right) \right] \quad [7.47]$$

The different parameters in [7.47] can be determined using the linear representation of Figure 7.13.

7.1.3.2. *Fed-batch reactor*

Fed-batch is effective in microbial reactors where the control of the substrate concentration is used to guide the reaction (e.g. production of baker's yeast where the sugar content in the substrate is controlled to limit ethanol production). This method involves initiating growth with a starter culture in volume V_0 until the substrate concentration at the maximum growth rate has been obtained (Figure 7.14). At this point, the batch is continuously fed at rate $\dot{V}$ so that the supply of substrate is equivalent to its consumption by microorganisms; it will therefore always remain at substrate concentration S, yielding a maximum rate. If the initial concentration of microorganisms X_0 is negligible compared to the microbial population $X = \psi\ (S_0 - S)$ generated in the reactor, the substrate concentration corresponding to the maximum rate is:

$$S = \sqrt{K_S^2 + K_S\ S_0} - K_S \qquad [7.48]$$

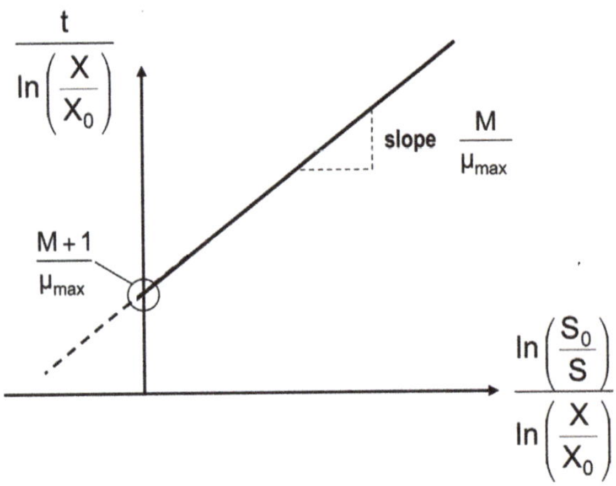

Figure 7.13. *Determination of microbial growth parameters*

If V_t and $\dot{V}_t$ are, respectively, the volume added and the feed rate at time t (Figure 7.14), the change in substrate concentration dS resulting from the addition of substrate at concentration S_0 for time dt is:

$$dS = \frac{\dot{V}_t\ (S_0 - S)\ dt}{V_0 + V_t} = \frac{\dot{V}_t\ (S_0 - S)\ dt}{V_0 + \int \dot{V}_t\ dt} \qquad [7.49]$$

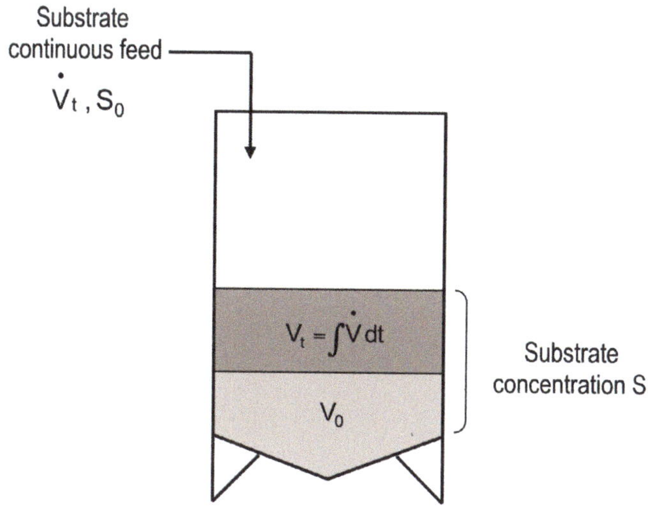

Figure 7.14. *Fed-batch reactor*

In order for the substrate concentration to remain equal to S, the substrate added should be consumed in dt according to $dS = \mu \frac{X}{\psi} dt = \mu \ (S_0 - S) \ dt$.

Hence:

$$\dot{V}_t = \mu \ (V_0 + \int \dot{V}_t \, dt) \qquad [7.50]$$

Therefore, the volume flow rate increases exponentially until the reactor is full: the culture develops according to the principles outlined in section 7.1.3.1.

7.1.3.3. *Continuous stirred tank reactor (CSTR)*

In the case of an enzyme reactor, the contents are well stirred and therefore homogeneous: the substrate and product concentration is constant throughout the reactor, and identical to that of the outgoing fluid. The enzyme should be retained within the reactor either by using enzymes immobilized on solid supports (Figure 7.15 (a)) or by coupling the reactor with an ultrafiltration membrane (Figure 7.15 (b)).

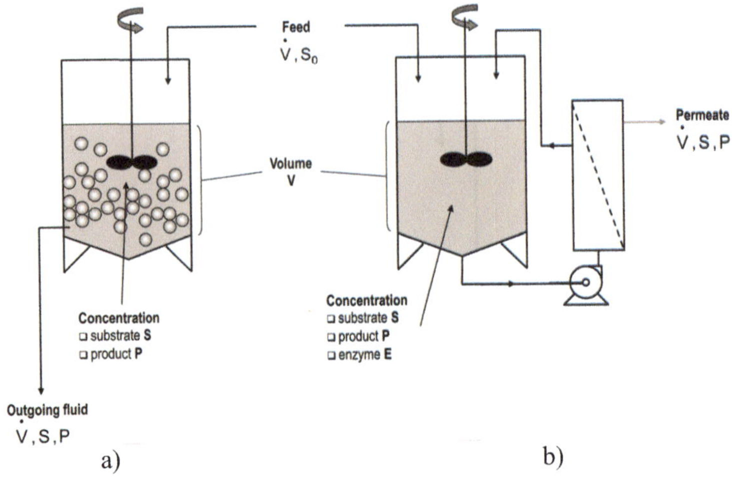

Figure 7.15. *Continuous stirred tank reactor: a) immobilized enzyme reactor; b) enzyme membrane reactor*

The modeling of these reactors is based on the Michaelis–Menten law. If $\dot{V}$ and V are, respectively, the volume flow rate and the volume of the reactor, the mass balance can be expressed as:

$$\dot{V}\,(S_0 - S) = V\,\frac{k\,E_t}{K_m + S}\,S \qquad [7.51]$$

where E_t is the total enzyme concentration. [7.51] expressed as a function of the conversion rate χ [7.40], gives:

$$K_m\,\frac{\chi}{1-\chi} + \chi\,S_0 = k\,E_t\,\frac{V}{\dot{V}} \qquad [7.52]$$

To determine the kinetic parameters (K_m and $v_{max} = k\,E_t$), [7.52] must be written as:

$$K_m + (1-\chi)\,S_0 = k\,E_t\,\frac{V}{\dot{V}}\,\frac{1-\chi}{\chi} \qquad [7.53]$$

and show $(1-\chi)\,S_0$ as a function of $\frac{V}{\dot{V}}\,\frac{1-\chi}{\chi}$.

In the case of a microbial bioreactor, microorganisms are totally or partially removed at the same time as metabolites and residual substrates; microorganisms are retained in the reactor using an ultrafiltration membrane (Figure 7.16).

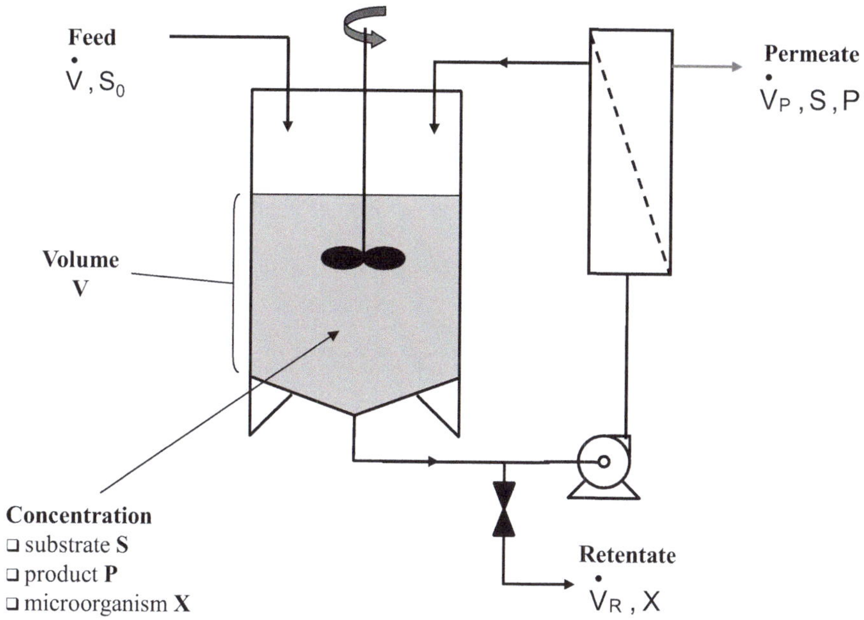

Figure 7.16. *Microbial membrane bioreactor*

For a reactor with volume V, fed at a flow rate $\dot{V}$ and a substrate concentration S_0 without biomass retention, the mass balance can be written as:

$$\dot{V}\ (S_0 - S) = V\ \frac{\mu_{max}\ S}{K_s + S}\ X = V\ \frac{\mu_{max}\ S}{K_s + S}\ \psi\ (S_0 - S) \qquad [7.54]$$

The residence time τ is equal to:

$$\tau = \frac{1}{\delta} = \frac{V}{\dot{V}} = \frac{1}{\mu_{max}} \frac{K_s + S}{S} \qquad [7.55]$$

where δ is the dilution rate in the reactor (s^{-1} or h^{-1}). The kinetic parameters can be determined by the linear representation of $\frac{1}{S}$ as a function of τ (Figure 7.17):

$$\frac{1}{S} = \frac{\mu_{max}}{K_S} \tau - \frac{1}{K_s} \qquad [7.56]$$

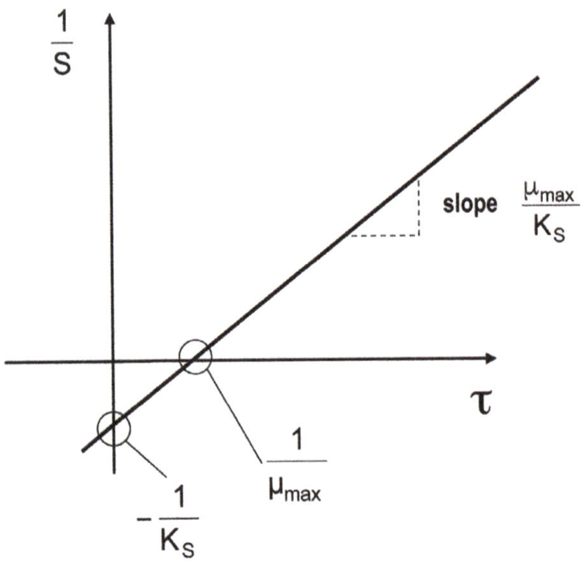

Figure 7.17. *Determination of kinetic parameters of microbial growth*

The change in substrate and cell concentration as a function of the dilution rate δ is shown in Figure 7.18. Beyond a certain value of δ, the microbial flora becomes depleted, the conversion rate drops and the amount of residual substrate increases. According to [7.56], $\frac{1}{S}$ rapidly decreases when $\delta = \frac{1}{\tau}$ tends towards μ_{max}.

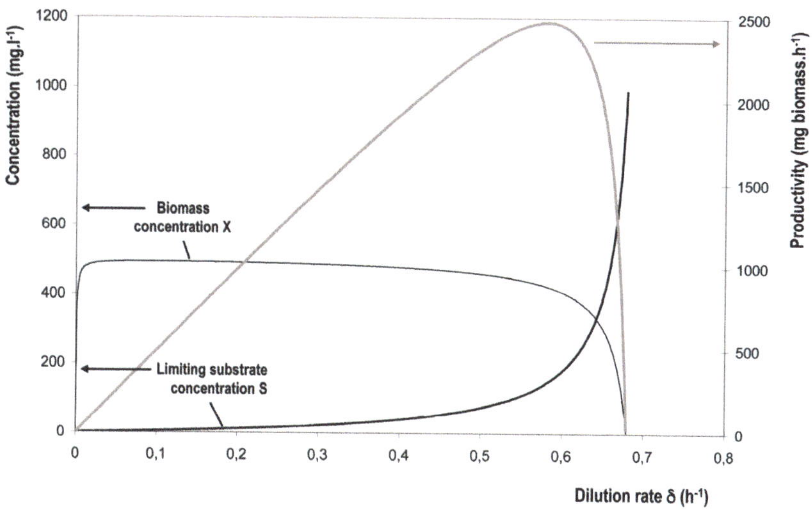

Figure 7.18. *Change in substrate concentration S (mg L^{-1}), biomass X (mg L^{-1}) and productivity as a function of the dilution rate δ (h^{-1})*

The reactor can operate at higher levels of biomass if the latter is partially retained (Figure 7.18), which increases productivity; this is useful when aiming at producing metabolites.

7.1.3.4. *Plug flow reactor (PFR)*

This type of reactor uses immobilized enzymes in a tubular chamber to create a fixed or packed bed through which 'piston' flow can be obtained, limiting the velocity gradients of the flow from the core to the wall (Figure 7.19). Biological reactions occur during the transfer: the progress of the reaction therefore varies between the inlet and outlet of the reactor.

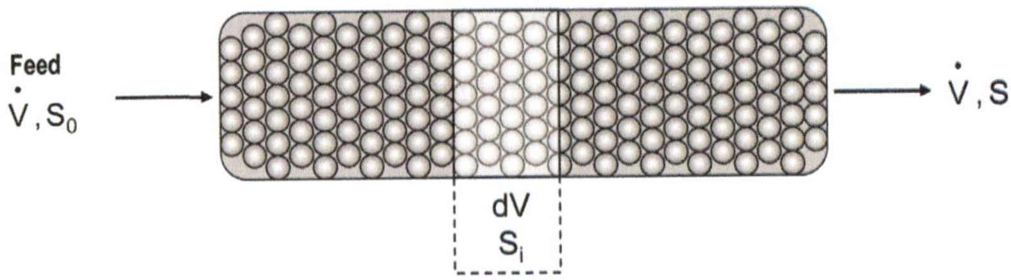

Figure 7.19. *Plug flow reactor*

The plug flow reactor can be modeled on the Michaelis–Menten equation, which is applied to a volume fraction of the reactor dV at a given point where the substrate concentration is S_i; if the reactor is fed at a volume flow rate $\dot{V}$, we can rewrite the equation locally for a volume fraction dV:

$$-\dot{V}\ dS = dV\ \frac{k\ E_t}{K_m + S_i}\ S_i \qquad [7.57]$$

Let:

$$-(K_m + S_i)\ \frac{dS}{S_i} = k\ E_t\ \frac{dV}{\dot{V}} \qquad [7.58]$$

By integrating between the inlet and the outlet of the reactor, we obtain a similar model to that described for the batch reactor with $t = \tau = \frac{V}{\dot{V}}$ [7.39]:

$$K_m\ \ln\left(\frac{S_0}{S}\right) + (S_0 - S) = k\ E_t\ \frac{V}{\dot{V}} \qquad [7.59]$$

This reactor can also be used with encapsulated microorganisms; since the microbial load is evenly distributed within the reactor and remains constant during the reaction (production of metabolites without the production of biomass), the model is similar to the previous one:

$$-\dot{V}\ dS = dV\ \frac{\mu_{max}\ S_i}{K_S + S_i}\ X \qquad [7.60]$$

By integrating, this results in an equation similar to [7.59]:

$$K_S\ \ln\left(\frac{S_0}{S}\right) + (S_0 - S) = \mu_{max}\ X\ \frac{V}{\dot{V}} \qquad [7.61]$$

7.1.3.5. *Plug flow reactor with recycle*

The performance of the reactor (volume V) can be improved by recycling part of the outflow at flow rate $\dot{V}_R$ according to Figure 7.20.

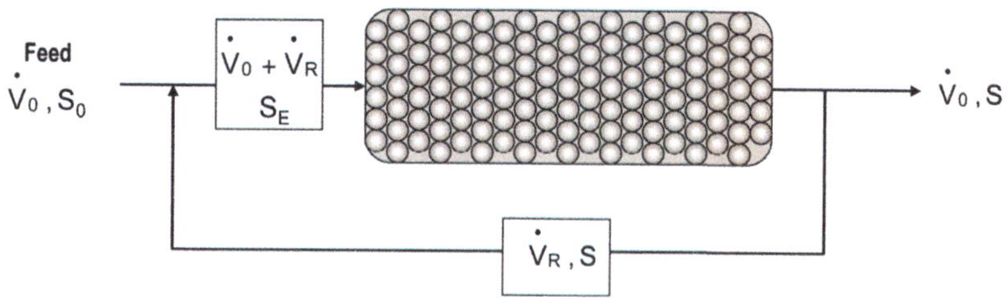

Figure 7.20. *Plug flow reactor with recycle*

The substrate concentration at the inlet of the reactor S_I is determined from the mass balance:

$$\dot{V}_0\ S_0 + \dot{V}_R\ S = (\dot{V}_0 + \dot{V}_R)\ S_I \qquad [7.62]$$

Therefore:

$$K_m\ \ln\left(\frac{S_I}{S}\right) + (S_I - S) = k\ E_t\ \frac{V}{\dot{V}_0 + \dot{V}_R} \qquad [7.63]$$

From [7.62], we can express S_I as a function of the recycling rate $\beta = \frac{\dot{V}_R}{\dot{V}_0}$:

$$S_I = \frac{\beta S + S_0}{\beta + 1} \qquad [7.64]$$

By replacing S_I by this expression in [7.63], we obtain:

$$(\beta + 1)\ K_m\ \ln\left(\frac{\frac{S_0}{\beta S} + 1}{\frac{1}{\beta} + 1}\right) + (S_0 - S) = k\,E_t\,\frac{V}{\dot{V}_0} \qquad [7.65]$$

In this last expression, $\frac{S_0}{\beta . S}$ and $\frac{1}{\beta}$ tend towards 0 as β increases. Under these conditions, $\ln\left(\frac{S_0}{\beta S} + 1\right)$ and $\ln\left(\frac{1}{\beta} + 1\right)$ tend respectively towards $\frac{S_0}{\beta S}$ and $\frac{1}{\beta}$, and the recycled model therefore becomes:

$$(\beta + 1)\ K_m\ \left(\frac{S_0}{\beta S} - \frac{1}{\beta}\right) + (S_0 - S) = k\ E_t\ \frac{V}{\dot{V}_0} \qquad [7.66]$$

As $\frac{S_0}{\beta S}$ and $\frac{1}{\beta}$ tend towards 0, this model is equivalent to that of a continuous stirred tank reactor (CSTR): a very high level of recycling in a plug flow reactor (PFR) therefore results in the equivalent of a stirred reactor.

7.1.4. *Criteria for choosing a bioreactor*

Several criteria apply in the selection of a bioreactor. Apart from batch reactors given their discontinuous nature, the three main selection criteria are:

– the desired conversion rate in an enzymatic reaction;

– inhibitions by reaction products or excess substrates;

– physicochemical changes to the reaction medium during the reaction.

Conversion rate

We have seen in CSTRs that concentrations of residual substrates and reaction products in the stirred reactor are those at the end of reaction; if the desired conversion rate is very high, the substrate concentration S at which the reaction takes place can be low, resulting in reaction rates well below v_{max} where S is less than or close to Michaelis–Menten constant K_m. In this case, the bioconversion potential of enzymes is underused.

However, the PFR is more efficient because a substrate concentration gradient exists between the inlet and the outlet of the reactor; thus, the reaction takes place at a maximum rate in the first part of the reactor where the substrate concentration is still high.

Inhibitions

When the reaction can be inhibited by the product, it is preferable to use the PFR because the product concentration is low at the top of the reactor and increases as the reaction progresses. However, a stirred reactor is more suitable if the reaction is inhibited by excess substrate since this type of reactor operates at the final concentration of substrate.

Physicochemical changes in the reaction medium

During the enzymatic reaction or the production of metabolites, changes in pH may occur (protein hydrolysis, lactic or acetic fermentation, etc.). This can have an adverse effect on the reaction rate constants, in which case it is necessary to adjust for pH changes through continuous pH measurements and the addition of alkali or acid. Such intervention poses no particular problem in a stirred reactor, but can be much more problematic in a fixed bed reactor.

Other phenomena can occur such as a change in the rheological properties of the reaction medium (increase in viscosity) or occasional precipitations. Such physicochemical changes can negatively affect fixed-bed reactors by creating pressure drops or clogging the reactor. It is therefore recommended to use stirred reactors.

Another key element is the cost of equipment and/or immobilized enzymes.

7.1.5. *Assembling bioreactors*

If stirred reactors (CSTR) are necessary and high conversion rates are desired, it is possible to place reactors in series; dividing reactors into a series of interconnected stages can significantly improve process performance.

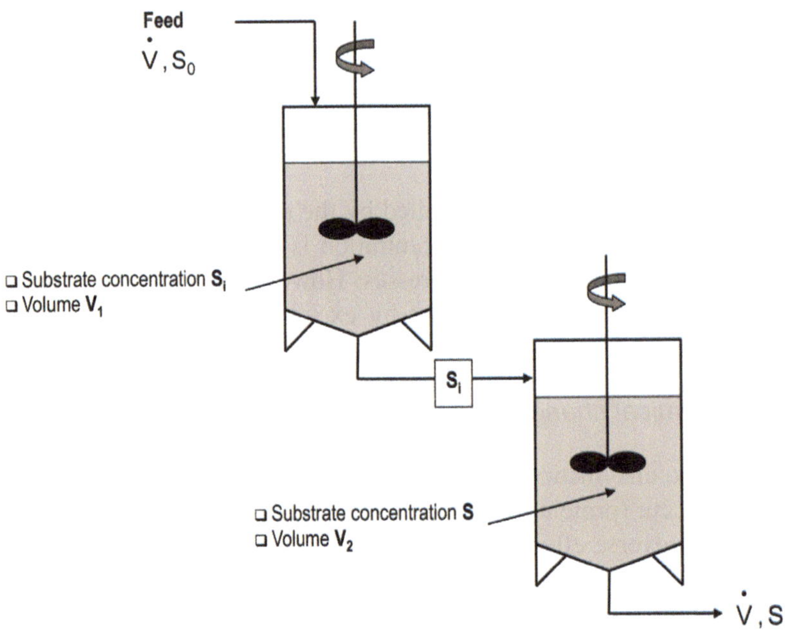

Figure 7.21. *Two CSTRs in series*

As shown in Figure 7.21, the first reactor operates at concentration S_i and the second at final concentration S. Thus, the first reactor can operate under

v_{max} conditions while the second runs at a rate below v; such a system is more efficient than one reactor with a volume equal to the sum of the two stages.

It is possible to show the improved performance graphically by dividing a CSTR into stages and comparing it to a PFR. In the case of a CSTR, we have:

$$\tau = \frac{V}{\dot{V}} = \frac{S_0 - S}{v_s} \qquad [7.67]$$

where v_s is the reaction rate at substrate concentration S. In the case of a PFR, the relationship is:

$$\tau = \int_{S_0}^{S} \frac{dS}{v_s} \qquad [7.68]$$

Plotting $\frac{1}{v}$ as a function of S (Figure 7.22) gives values of τ for each type of reactor:

– the residence time in the CSTR corresponds to $(S_0 - S) \times \frac{1}{v_s}$ (Figure 7.22, dark gray area A);

– the residence time in a two-stage CSTR is equal to the sum of τ_1 and τ_2, respectively corresponding to areas $(S_0 - S_i) \times \frac{1}{v_{si}}$ (light gray) and $(S_i - S) \times \frac{1}{v_s}$ (mid-gray) in Figure 7.22;

– the residence time in the PFR corresponds to the area under the curve of Figure 7.22.

When $\tau_{CSTR} > (\tau_1 + \tau_2)_{CSTR} > \tau_{PFR}$, and the greater the number of stages, the more the performance tends towards a PFR.

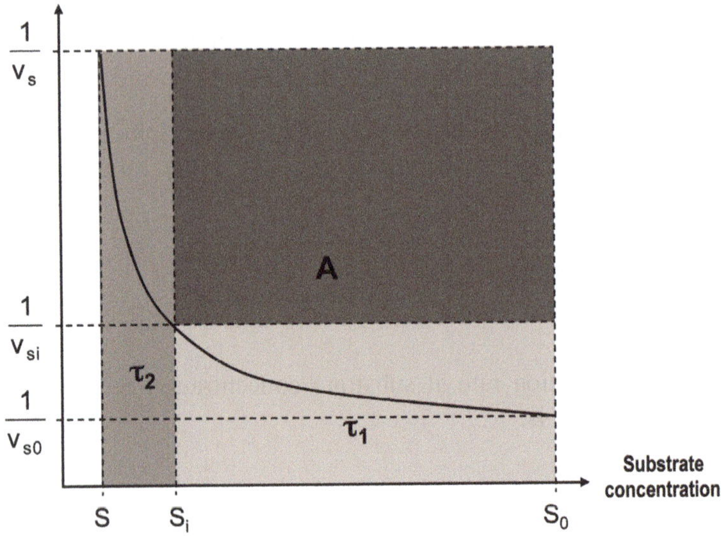

Figure 7.22. *Graphical representation of residence times in reactors*

In the case of two CSTRs in series, it is possible to calculate the substrate concentration S_i and consequently the residence times or volumes of each stage to obtain optimal conditions. To do so, it is necessary to obtain the shortest possible residence time, corresponding to the largest possible contact area:

$$A = (S_0 - S_i)\left[\frac{1}{V_s} - \frac{1}{V_{si}}\right] \qquad [7.69]$$

The derivative of A with respect to S_i is:

$$A' = \frac{K_m}{k\ E_t}\left[\frac{S_0}{S_i^2} - \frac{1}{S}\right] \qquad [7.70]$$

This derivative is zero for $S_i^2 = S_0\,S$. If the enzyme concentration is the same in both reactors, it is possible to determine the residence times and volumes of each stage, knowing S_i and that the flow rate is the same.

Combining a PFR and a CSTR is also possible as shown in Figure 7.23.

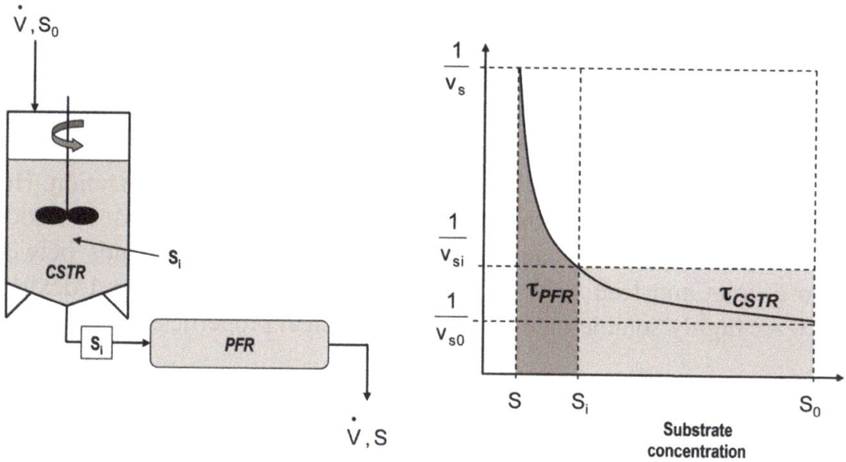

Figure 7.23. *CSTR–PFR combination*

7.2. Physicochemical changes

Since the techno-functionality of biomolecules is closely dependent on their structure and physicochemical properties, any change induced by thermal or chemical treatment could generate new functionalities that better respond to the technological requirements of the secondary processing industry. One of the main constraints is stability with regard to thermal (sterilization–freezing–thawing), hydrodynamic (shear during transfers) and mechanical treatments (pumping). The main principles of physical-chemical changes are not covered here as this is dealt with in Volume 3 [JEA 16b].

7.2.1. *Heat and mechanical treatment*

These treatments mainly concern carbohydrates and proteins.

In the case of polysaccharides such as starch, heat treatment has a triple effect with significant consequences in terms of functionality: disruption to the crystalline structure, hydration and hydrolysis. As already outlined in Volume 1, starch consists of two types of subunits, amylose and amylopectin, which are chains of glucose units linked by α(1-4) glycosidic bonds for amylose and α(1-4) and α(1-6) glycosidic bonds for amylopectin. Native starches are semi-crystalline in structure and are stabilized by a large number of OH groups thus limiting their hydration, thickening and gelling power. It also makes them less susceptible to enzymatic attack, and therefore limits digestibility. The heat treatment of aqueous starch suspensions disrupts the crystalline structure, which induces a release of OH groups, making it possible to hydrate. This results in a swelling of the grain and an increase in viscosity of the dispersed system, known as gelatinization. Heat treatments, depending on their intensity and physicochemical conditions (water content, pH) can cause the hydrolysis of covalent bonds between glucose units and lead to the formation of glucose, maltose and dextrins, which are highly soluble but have poor rheological properties.

As the gelatinized starch dispersion (starch paste) cools, recrystallization can occur, known as retrogradation, with major consequences from a rheological perspective. This phenomenon increases as cooling slows in the temperature zone above the glass transition temperature, which largely depends on the water content (Figure 7.24). The functional properties therefore depend on the type of starch (amylose/amylopectin ratio), the intensity of the heat treatment (time and temperature), the heating and cooling kinetics as well as the water content. A shear effect can be combined with hydrothermal action in extrusion; thus treatments of 20–60 seconds in the presence of 10 to 40% water at temperatures above 200°C and pressures of 20 MPa are commonly used to prepare ingredients intended for use in sauces, soups, beverages, etc.

Proteins are also highly sensitive to heat treatment; as the bonds in the secondary and tertiary structures of proteins are generally low energy (ionic, hydrogen, van der Waals), heat treatment can disrupt them and consequently induce new structures and functionalities. Structural disorganization can cause an increase in surface hydrophobicity and greater reactivity by exposing the hydrophobic side chains of amino acids and thiol groups. These denatured proteins exhibit new physicochemical and functional properties such as the ability to aggregate or gel at room temperature under certain ionic strength and pH conditions and stabilize dispersed emulsion or foam systems.

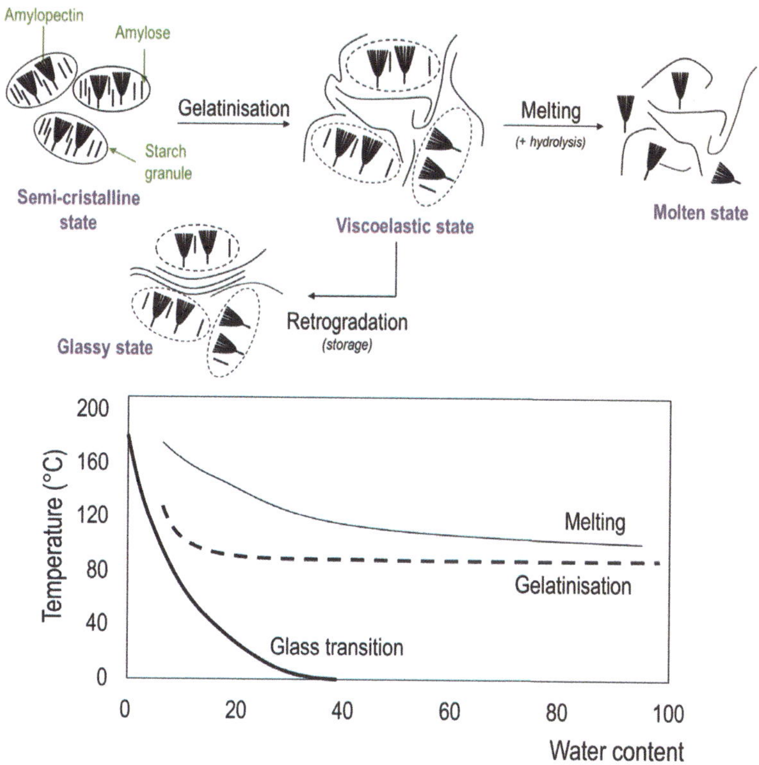

Figure 7.24. *Structural changes in starch as a function of temperature*

Controlling the thermally induced functionalization of proteins requires a good knowledge of the different stages of denaturation and their kinetics (Figure 7.25). Modifications to the secondary and tertiary structure generally follow first order kinetics whereas aggregation or polymerization follow second order kinetics, and are highly dependent on the physicochemical conditions of the medium (pH, ionic strength, type of electrolytes, redox potential, etc.). The effect of heat treatment on protein functionality therefore depends on the intensity of the treatment (time and temperature), the protein concentration and the ionic environment. As with starches, subjecting native or denatured proteins to high pressure shear treatment sometimes combined with heat treatment is used to prepare a wide range of ingredients used as cheese, meat and fish substitutes.

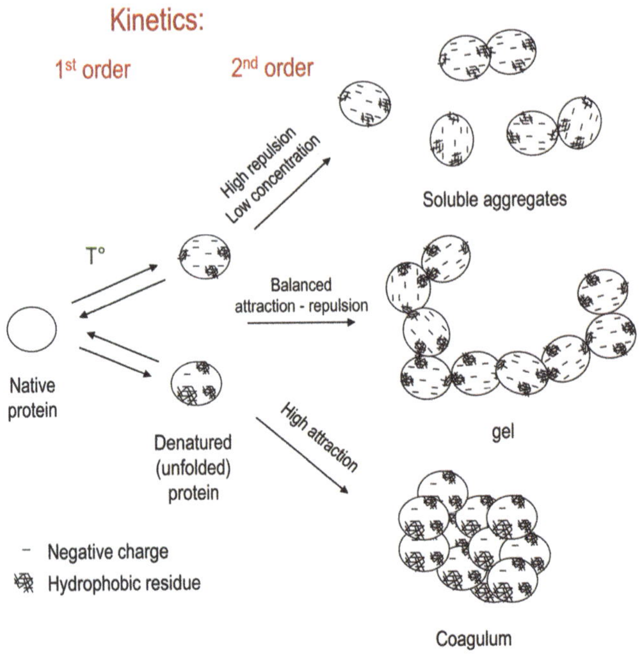

Figure 7.25. *Protein denaturation/aggregation as a function of the treatment*

7.2.2. Crosslinking of macromolecules

The crosslinking of macromolecules involves the formation of new intramolecular or intermolecular bonds, which creates new rheological properties or improves their stability to processing such as high temperature treatment (sterilization) or medium conditions (pH) that could cause the rupture of covalent bonds. Crosslinking can be carried out either by exploiting the reactivity of functional groups in the macromolecule or by chemical reagents capable of interacting with a functional group in the macromolecule.

With starches, crosslinking is obtained using bifunctional reagents capable of reacting with two hydroxyl groups; these reagents may be chloroepoxides, phosphate derivatives, or diacid anhydrides that can form intra- or intermolecular bonds such as those shown in Figure 7.26.

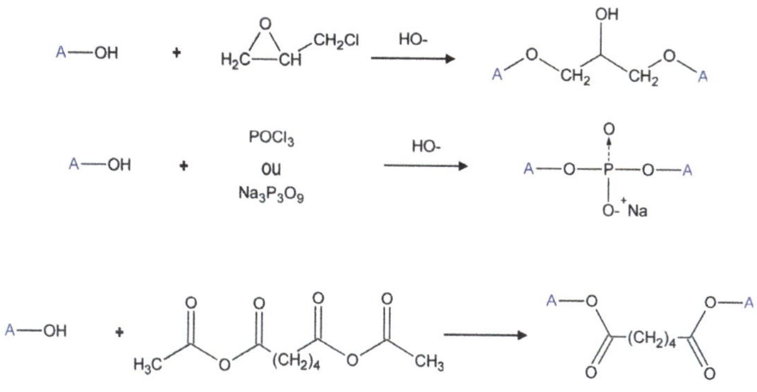

Figure 7.26. *Crosslinking of polysaccharides*

In the case of proteins, the reactivity of amino acid side chains offers different crosslinking possibilities (Figure 7.27):

– amide (glutamine–asparagine) and carboxyl groups (glutamic or aspartic acid) can interact with the ε-NH_2 of lysine to produce isopeptides;

– serine (phosphoserine) or cysteine can lead to the formation of dehydroalanine by β elimination, which reacts with basic amino acids (lysine, arginine) or cysteine to give lysinoalanine, ornithoalanine and lanthionine;

– cysteines can form disulphide bonds by oxidation or form new disulphide bonds and thiol groups by the nucleophilic attack of thiol/thiolate groups in disulphide bonds;

– tyrosine can dimerize by the condensation of two phenol groups.

These types of crosslinking are favored, under oxidative conditions or during heat treatment, when the amino groups are not protonated and the thiol groups are in the form of thiolates: they are generally faster at slightly alkaline pH. This type of interaction can generate new functionalities, but has some nutritional disadvantages since the new bonds formed are not always biodegradable. Crosslinking is applied to food products with a pH slightly above neutral, especially during heat treatment.

In the case of proteins, it is possible to add functional groups to improve their solubility and/or interfacial properties; a reaction that is easy to implement is based on the reactivity of amines (ε-NH_2 group of lysine) towards reducing carbohydrates (glucose, lactose, maltose, maltodextrin, etc.).

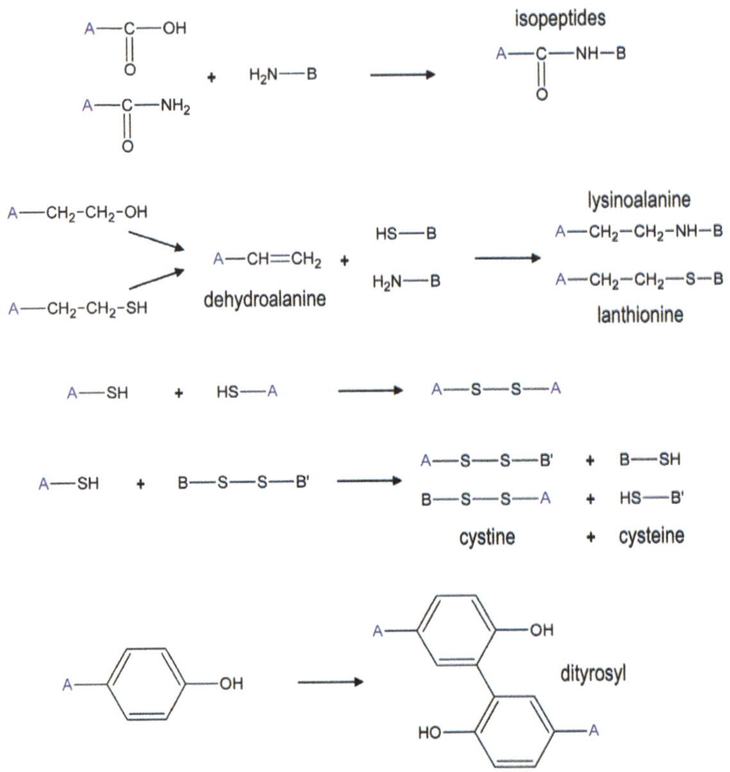

Figure 7.27. *Crosslinking of proteins by the interaction of amino acid side chains*

7.2.3. *Addition of functional groups*

The addition or grafting of anionic or cationic, hydrophilic or hydrophobic groups onto polymers can either change their functional properties or improve their stability to thermal and/or shear stress (Figure 7.28).

For instance, acetylation and alkylation can stabilize starches against cold temperatures by limiting retrogradation.

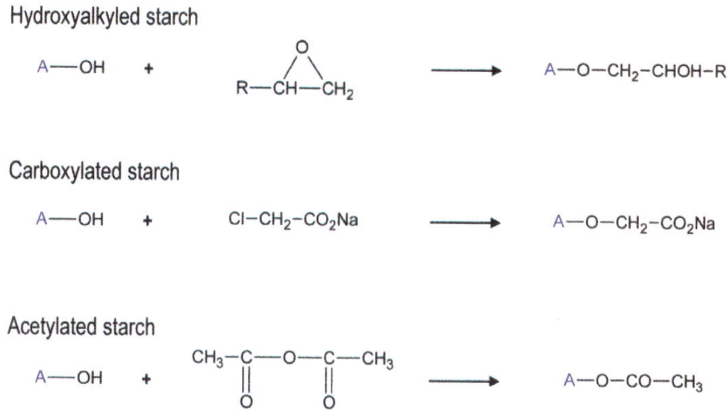

Figure 7.28. *Addition of functional groups onto starch*

7.2.4. Hydrogenation

The hydrogenation of biconstituents mainly concerns two types of biomolecules: reducing carbohydrates and unsaturated fatty acids.

The chemical reduction of carbonyl groups of carbohydrates (glucose, galactose, fructose, xylose, maltose, lactose, etc.) is used to prepare polyols (sorbitol, mannitol, xylitol, lactitol, maltitol, etc.), which have the advantage of being non-cariogenic (does not promote tooth decay) and non-insulinogenic, and are consequently used in diet products.

The chemical hydrogenation of saturated fatty acids alters the physicochemical properties of fats by increasing the melting temperature and reducing sensitivity to oxygen. These transformations are carried out in high pressure hydrogen reactors in the presence of catalysts (copper, nickel) at temperatures of up to 200°C.

PART 5

Packaging

8

Packaging: Principles and Technology

Packaging is the final operation in the production of food. It is inseparable from the product and should help to preserve hygienic, nutritional and sensory qualities, meet logistical and distribution constraints and satisfy consumer expectations. Packaging is also a source of information and communication, which can significantly influence product perception and purchasing decisions. Technological advances in packaging over the past 20 years have led to innovation in terms of convenience and extended shelf life (longer use-by and best-before dates) by ensuring better quality during storage.

Packaging food may involve filling, counting or weighing, depending on the product. It can be carried out under aseptic conditions depending on the stabilization process applied after packaging (canning, high pressure treatment, ionization). Aseptic packaging is applicable to fresh products or previously pasteurized or sterilized products in bulk (UHT); it reduces microbial contamination by the sterile filtration of air, or limits microbial growth and certain chemical reactions detrimental to sensory quality by modifying the atmosphere during the process. Products that are heat-treated in bulk and then packaged under aseptic conditions generally have better sensory and nutritional quality than products that have undergone heat treatment after packaging. The packaging used can be produced externally or on-site.

Chapter written by Valérie LECHEVALIER.

Vacuum packaging and modified atmosphere packaging are very common in the food sector as they are well suited to market trends (extending the shelf life of fresh products, improving quality); furthermore, progress in the area of packaging materials has improved their convenience (resealable film, steam vents, heat-proof film, etc.).

8.1. Packaging: definition and principles

European directive 94/62/EC gives the official definition of packaging and its field of application. It refers to packaging as meaning "all products made of any materials of any nature to be used for the containment, protection, handling, delivery and presentation of goods, from raw materials to processed goods, from the producer to the user or the consumer". Packaging consists of:

– *sales packaging* or *primary packaging*, that is packaging conceived so as to constitute a sales unit to the final user or consumer at the point of purchase;

– *grouped packaging* or *secondary packaging* consists of a certain number of sales units intended for the final user or consumer; it can be removed from the product without affecting its characteristics;

– *transport packaging* or *tertiary packaging*, that is packaging conceived so as to facilitate handling and transport of a number of sales units or grouped packagings in order to prevent physical handling and transport damage.

Materials intended to come into contact with food (food contact materials or FCM) must comply with the requirements set out in Regulation 1935/2004 EC:

– food contact materials must not transfer their components into food in quantities that could endanger human health, change food composition in an unacceptable way or deteriorate its taste, texture or odor (in line with the principle of inertia for FCM);

– food contact materials or constituents thereof must be included in lists of authorized substances (in line with the principle of composition);

– labeling, advertising and presentation of food contact materials must be explanatory and not mislead consumers.

8.2. Functions of packaging

The functions of packaging must meet the requirements of all the participants in the product distribution chain from the factory to the consumer, for example, an easy-to-open package for the consumer must be strong enough to withstand the mechanical stress of transport and shelving. Packaging design can therefore be a challenging task and must integrate a wide range of functions [MUL 98, EMB 12a].

8.2.1. *Technical functions of packaging*

8.2.1.1. *Containment*

The primary function of packaging is the containment of the product. Packaging a product requires knowledge of its composition and physical properties to identify any storage limitations: a liquid or solid are not packaged in the same way. The choice is therefore based on a "product–packaging" combination. It is also necessary to take into account the amount of product to be packaged, which means considering the combination "product–packaging–conditions of use". This element is constantly changing depending on consumer lifestyles, with a tendency in recent years towards individual or daily consumption units.

8.2.1.2. *Logistics*

The logistics function of packaging must meet transport and storage requirements and facilitate product handling. From an ergonomic point of view, the ideal scenario would be to have a base unit 40 cm wide (corresponding to the width between the shoulders) and to package everything in multiples of this unit (boxes, palettes, trucks corresponding to a multiple of palette widths).

8.2.1.3. *Protection*

The main function of packaging is the preservation and protection of the food product together with an obligation to provide toxicological safety and chemical inertness of packaging materials. Packaging is primarily a barrier between the product and the external environment, thus ensuring the *passive protection* of the product. This can be mechanical, protecting against potential shock and stress, but also against insects and rodents. It must also protect the product from mass transfer that can occur in the liquid phase (impermeability or porosity of packaging of liquids) or gas phase (tightness or porosity to gas and other volatile substances). Packaging permeability is dealt with further on in section 8.3. Mass transfer can be:

– *from the external environment to the product*: packaging acts as a water barrier to prevent mould or a deterioration in texture, an oxygen barrier to avoid aerobic bacterial growth and oxidation, and a barrier to volatile substances in the environment (hydrocarbons, smoke, scents, etc.), which may alter the organoleptic properties of the food;

– *from the product to the external environment*: packaging acts as a barrier to water therefore avoiding dehydration of the product, a barrier to gas or gas mixtures (CO_2, N_2, etc.) that may be included in the packaging to ensure product preservation, and a barrier to volatile compounds to prevent flavor loss.

Packaging should also provide protection against energy transfer by radiation (light) or conduction/convection (heat), which may occur from the external environment to the product and initiate or accelerate microbiological spoilage or chemical processes. Many food products are sensitive to light (visible, near infrared or ultraviolet), which triggers photochemical reactions causing discoloration, vitamin loss and amino acid photolysis. For such products, the photo-protective role of packaging is to filter the relevant wavelengths or even stop any light entering (opaque packaging). As regards heat transfer, the insulating nature of packaging is useful whenever the product-package is subjected to temperature changes.

Finally, packaging also acts as a barrier between microorganisms in the external environment and the product. It therefore plays a critical role in maintaining the hygienic and microbiological quality of food by forming a physical barrier between microorganisms and the product thereby preventing post contamination, as well as preventing or limiting mass transfer (water, gas) that could promote the growth of any germs if present. In addition, it is important to remember that in the case of sterilized products, packaging must be able to withstand the time/temperature ratio applied to aseptic packaging.

A new generation of packaging has been developed over the past 30 years, known as active packaging, that is it has active functions beyond the inert passive containment and protection of the product. Active packaging interacts with the product and in some cases responds to changes in the surrounding environment or the product itself [WAG 89]. It therefore exploits possible interactions between the packaging and the food product whereas up until now these interactions occurred unintentionally and were kept to a minimum due to existing legislation.

In terms of quality control, this new generation of packaging can be divided into two main categories: *active*, which acts on the food or its environment to preserve quality throughout the distribution chain, and *intelligent*, which has the ability to sense or measure storage conditions or food properties and can inform the consumer in real time [DAI 08]. Only active packaging may be considered as having a food protection function. All food spoilage mechanisms are targeted by the action of such packaging: lipid oxidation, browning, degradation of vitamins, pigments and flavors, microbial growth, respiratory activity of fresh vegetables, etc. They act either by being in direct contact with the food (ionomeric films onto which antimicrobial agents are grafted), by ensuring the optimal gas composition of the internal atmosphere of the package through adsorption (oxygen, ethylene, water vapor, aromatic compounds), by salting out (carbon dioxide, ethanol, sulfites, antimicrobial agents (sorbate, lysozyme) or antioxidants, vitamin E, butylatedhydroxytoluene) or by the selective transfer of gas between the internal atmosphere and the surrounding environment. The latter can be considered as both conventional and active packaging.

The protective function of packaging is therefore threefold since it protects the product from the external environment as well as from itself (e.g. respiration of vegetables) and also protects the external environment from the product (e.g. release of flavors).

As regards intelligent packaging, the best known and most widely used system is a time/temperature indicator, whereby the color changes based on the storage or processing conditions of the product. Some indicators take into account the overall processing history of the product whereas others only indicate that a critical temperature has been exceeded (break in the cold chain). There are also oxygen and carbon dioxide indicators in the internal atmosphere, which ensure package integrity. Some indicators go further by detecting certain volatile compounds involved in the freshness of the product such as volatile amines in fish, while others allow you to control the production process directly from within the package (e.g. monitoring pH during fermentation).

8.2.1.4. *Utility of use*

This is a challenging function in the sector. It must be present from the packaging stage (automated packaging, storage and simplified delivery, limited loss, etc.) to delivery to the consumer (guaranteed quantities and consumption units, ease of handling and storage, ease of opening and safety in use, etc.) via the distributer (stock control, optimization of retail space and shelving, traceability, optimization of check-out time, etc.). It provides the user with information about the product itself or storage conditions. In order to meet the requirements of traceability and information, the barcode may soon be replaced by radio-frequency identification (RFID). Packages would then contain a microchip containing non-modifiable information that must by law be displayed on packaging as well as information that can be modified by the distributer (price, anti-theft labeling, etc.). This system could also help to accelerate the check-out process.

Packaging can also be used to prepare food before consumption. It can modify the characteristics of food, mainly temperature, to facilitate consumption (self-heating or self-cooling packaging). Water-soluble and edible packaging is also being developed for portions of soups or sauces in Japan, for example.

8.2.2. *Communicative functions of packaging*

8.2.2.1. *Marketing*

Since packaging cannot be disassociated from the food product, it significantly influences the purchasing decision of the consumer. It should attract attention, arouse interest and create a desire to purchase. This purely marketing function can be broken down into five sub-functions:

– *alert* corresponds to the impact of packaging on the consumer via a strong signal often linked to a brand. Several elements can be used such as graphics, shape, color or material, with the key being that the product stands out;

– *assignment* allows customers to place the product in their world of reference. Codes exist in terms of color, graphics and materials, for example, the packaging of chocolate bars is often brown or black. Manufacturers can either adhere to or disregard the world of reference;

– *positioning* specifies a particular place within a product range (e.g. budget, mid-range, luxury, etc.);

– *seduction* is the most subjective function of packaging; it is related to aesthetics, shape, color and the symbolism of packaging materials, which should trigger purchase. Re-purchase of the same product requires a good packaging-product match;

– *loyalty* corresponds to the aim of creating a link between the consumer and the product. Packaging should highlight the specific features of the contents and prevent the purchase of a competing product.

8.2.2.2. *Information*

Packaging is the best medium for providing information on the product. A distinction is made between mandatory information, as defined by law, and useful information.

Mandatory information is defined by the directive 2000/13/CE recently complete by the INCO regulation (no.1169/2011), which requires a description of the foodstuff (legal trade name, list of ingredients, net quantity, nutritional values, use by or best before date, identification of sales agent, production batch and other additional information specific to certain foods, such as the percentage of alcohol in alcoholic beverages for example), as well as the directive 98/6/CE on the Freedom of Prices and Competition, which requires the display of price. The directive 2000/13/CE also specifies that labeling and the manner in which it is carried out should not mislead consumers.

Useful information is at the discretion of the producer. It may include instructions or suggestions for use, reminders of other products in the range, quality signs (controlled designation of origin (French AOC), protected designation of origin (PDO), protected geographical indication (PGI), agricultural labels, certifications of compliance, organic agriculture labeling, etc.), apparent benefits from consuming the product, symbols of the packaging materials or the Green Dot certifying that the manufacturer of the product has paid environmental tax. Some of this information is subject to legislation.

8.2.2.3. *Communication*

Packaging is both the subject and medium of communication. Manufacturers sometimes advertise and customize their products via packaging (e.g. quick cooking boil-in-the-bag rice, microwaveable dinners, etc.). Packaging is also an excellent medium of communication because it is seen and read by millions of consumers; thus it is possible to target consumers.

8.2.3. *Environmental function of packaging*

Apart from the aforementioned functions, packaging should be compatible with the environment. Eco-design, which aims to decrease the environmental impact of packaging, must be taken into account at all times. It was formalized by European Directive 94/62/EC, which states that all packaging placed on the market since 1 January 2000 must meet key requirements, which are:

– *reduction at source* by reducing packaging volume and weight through modifying the characteristics of the product (denser, smaller, more concentrated, etc.), the packaging process (compression, reduction of empty space, etc.) or packaging design (eco-refills, simplified packaging, size optimization, type and implementation of materials used);

– *reduction of landfill disposal* through reuse (deposit, etc.), recycling, composting and/or energy recovery (incineration).

This directive also sets target figures for the recovery of different types of packaging (wood, plastic, metal, glass and cardboard), and in particular recycling [EMB 12b].

According to the "polluter pays" principle, manufacturers are required by law to ensure the disposal of their packaging. As it is physically impossible for a manufacturer to recover empty packaging from individuals, there are local and national organizations and companies that deal with this issue. In principle, manufacturers have the option to join these organizations, but in practice it is difficult not to. They can then display the Green Dot on their packaging (Figure 8.1).

Figure 8.1. *The Green Dot symbol*

8.3. Properties of packaging material

The characteristics of materials used in packaging determine its functions and should therefore be used to slow down the physicochemical and microbial changes in the food product. A number of measurable characteristics exist (a_w, peroxide index, color, aromatic composition, texture, etc.) for which it is possible to determine a limit value; beyond this limit, the product is considered non-compliant. In addition, the product can also be an excellent nutrient medium for microorganisms.

The properties of packaging materials, in terms of their functions, are summarized in Figure 8.2. It is important to consider the contact time between the product and the material, which depends on mass transfer, as well as the contact time between the external atmosphere, the material and the internal atmosphere, which depends on the permeability of the packaging.

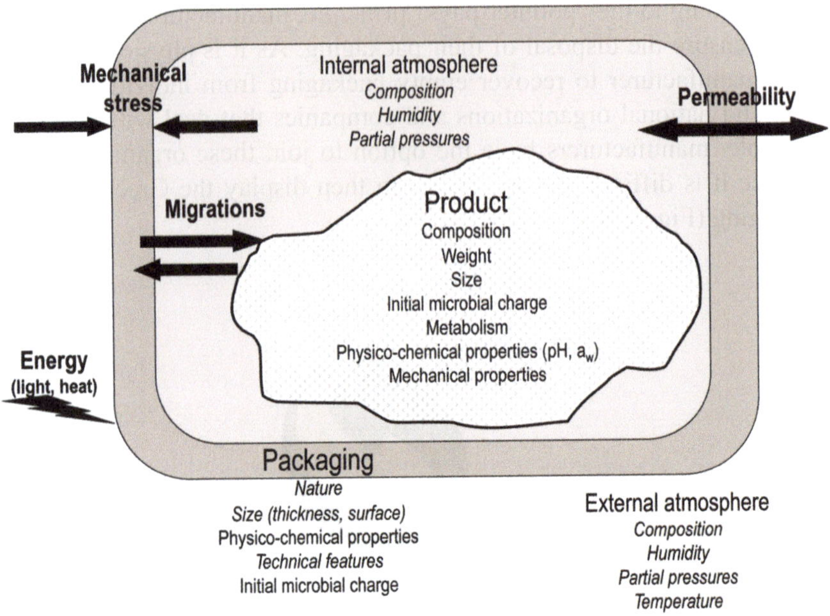

Figure 8.2. *Properties of packaging materials*

8.3.1. *Permeability*

Permeability is a defining characteristic of packaging and determines the mass transfer through the packaging material. These bidirectional transfers (from the external to the internal atmosphere and vice versa) can occur in the liquid phase, or more commonly in the gas phase (gas, vapors and other volatile substances).

8.3.1.1. *Physical principles*

The transfer of water vapor or gas through a material is described by the theory of permeation, which involves a transfer by molecular diffusion induced by a difference in concentration of the diffusing gas on either side of the material. Permeation is often described as a succession of three steps: the penetrant adsorbs onto the material, diffuses through the material under the effect of a concentration gradient and desorbs by evaporation. In the steady state, for a constant temperature and partial pressure gradient and an inert and isotropic material, unidirectional diffusive mass transfer is described by Fick's first law:

$$\varphi_m = -D_m \frac{dC}{dx} \tag{8.1}$$

where ϕ_m is the mass flux density of the penetrant, D_m is the diffusivity of the material (considered constant for the whole material) and $\frac{dC}{dx}$ is the concentration gradient. Sorption (adsorption/desorption of the penetrant from the surface to the material) depends on the solubility of the diffusing substance in the material. At constant temperature and equilibrium, Henry's law is used to express the concentration of the penetrant (C):

$$C = S\,P \tag{8.2}$$

where S is the solubility coefficient and P is the partial pressure. By combining [8.1] and [8.2], we obtain:

$$\varphi_m = -D_m\, S \frac{\Delta P}{\Delta x} \tag{8.3}$$

where ΔP is the difference in the partial pressure of the penetrant on either side of the packaging. The product of the diffusion coefficient (D_m) and the solubility coefficient (S) is the permeability coefficient or the permeability of the material (P_e), which can be linked to the flux of the penetrant in the steady state:

$$P_e = \phi_m \frac{\Delta x}{\Delta P} \tag{8.4}$$

The permeability coefficient is only used when permeability is constant regardless of the pressure gradient, where D_m and S are independent of the penetrant concentration. In practice, there are many cases where the penetrant interacts with the material, which leads to the formation of concentration gradients within the material. Diffusion and solubility coefficients increase with the difference in partial pressure due to the affinity of the material for the permeant. In this case, permeability can be calculated using the following equation:

$$P_e = \frac{m\ x}{A\ t\ \Delta P} \qquad [8.5]$$

where m is the amount of permeant (in kg, mol or m^3), x is the thickness of the material (m), A is the surface area of the material (m^2), t is time (s) and ΔP is the partial pressure difference (Pa). Gas and water vapor permeability are measured under variable conditions of temperature, relative humidity and thickness of the material, which makes it difficult to compare different materials [JAS 94].

8.3.1.2. *Parameters affecting permeability*

The phenomena of diffusion and solubility that define permeability depend on the physical properties of the material, the penetrant and the medium conditions.

Properties of the material

For simplicity, it is possible to equate the structure of a material to that of a sieve. If the mesh is small and it is difficult to change its dimensions, the energy required for the transfer of the diffusing material is high and therefore permeability is low. No material, except for a welded metal box or a glass jar, is absolutely gas-tight. A suitable choice must therefore be made between the types of materials or combinations thereof to optimize permeability with regard to different gases, depending on the type of product, preservation method, storage conditions, environmental pollution risks and shelf life.

Properties of the permeant

The dimensions, configuration, polarity and condensation capacity of the diffusing gas all affect permeability. The size of the penetrating gas molecules mainly influences diffusion and solubility coefficients. For a given material, diffusion decreases as the diameter of the diffusing compound increases. However, the solubility of a material depends on the structure of the diffusing compound and this dependence follows a linear relationship depending on the molar volume. In addition, interactions can occur between the packaging and the penetrant based on their respective polarity, thus significantly altering the permeability of the penetrant or other molecules.

Water is a special case since water barrier properties are defined depending on its state: impermeability to water in the liquid state, evaporation within the material and the diffusion of water vapor through the walls of the container.

Environmental conditions

The diffusion coefficient, solubility and consequently the permeability of a material vary with temperature according to the Arrhenius law:

$$D_m = D_0 \, e^{-\frac{E_d}{RT}} \qquad [8.6]$$

$$S = S_0 \, e^{-\frac{E_s}{RT}} \qquad [8.7]$$

where D_0 and S_0 are constants, E_d and E_s are activation energies, R is the ideal gas constant and T is absolute temperature. Assuming Henry's law applies, we obtain:

$$P_e = P_0 \, e^{-\frac{E_p}{RT}} \qquad [8.8]$$

Diffusion always increases with temperature; with solubility, on the other hand, the activation energy depends on the type of diffusing gas. In most food packaging, solubility only depends to a small extent on temperature and therefore the activation energy remains positive. Hence, permeability increases with temperature. However, when the gas interacts with the material, as in the case of organic vapors for which the solubility and diffusion coefficients depend on their concentration in the material, [8.8] no longer applies. Permeability therefore increases with temperature and the concentration of the scattering material in the polymer. An inversion of the slope can be observed above a certain temperature threshold. Finally, some materials undergo structural changes at a precise temperature, which may affect the permeability coefficient.

For simple gases (oxygen, carbon dioxide, nitrogen), the transfer rate through the material is directly proportional to the partial pressure gradient: thus, in the case of a food product packaged under nitrogen, the transfer of oxygen through the packaging occurs when the external partial pressure of the gas exceeds the internal partial pressure. In the case of water vapor, two scenarios are possible: the vapor does not dissolve in the material and the transfer rate is proportional to the difference in partial pressure on either side of the material at equilibrium, or the water vapor dissolves in the material and the transfer rate consequently depends on the partial pressures and absolute pressure.

8.3.1.3. *Selective permeability*

Packaging materials, particularly polymers, can be more permeable to certain compounds than others. This selectivity is specific to dense materials in which the permeability mechanisms occur by the solubilization and diffusion of the penetrant. For most polymers, permeability to carbon dioxide is about three times greater than that of oxygen, which is itself six to eight times greater than nitrogen. Gas barrier properties should therefore be evaluated case by case, depending on the type of gas, the type and thickness of the given material, the desired shelf life and the required gas pressure in the container.

There is therefore no general solution for packaging permeability, but rather specific solutions to each particular case. In all cases, it is essential to have a good knowledge of the product, its development over time and its sensitivity to degradation reactions.

8.3.2. *Migration*

Migration corresponds to the transfer of substances from the packaging material to the product by physicochemical reactions, except for elements that could be separated from the packaging due to mechanical stress. A distinction is made between specific migrations, where the transfer of identified components is measured, and global migration, where all the components of the packaging material that have come into contact with the product are measured [CAS 07].

8.3.2.1. *Potential migrants*

It is important to identify and understand the potential migrants, especially those with legal limits. These migrants can have several sources: they may be present in the raw material used to make the packaging (metal, wood tannins, sand in glass, residues from the synthesis of monomers, substances incorporated during recycling, etc.), they may have been incorporated intentionally or unintentionally during the processing of the raw material (monomers, additives, catalysts, degradation products of these additives and polymers, etc.) or they may appear during storage (reactivity of additives, degradation of material, reactivity between food constituents and the container) or heating of the packaged product (microwaving).

These migrants can be classified according to their molecular weight, which affects their ability to migrate. Most migrants are easily detected by chromatographic and spectrophotometric methods (Table 8.1).

Molecular weight (Da)	*Characteristics*	*Analysis technique*
from 40 to 150–200	Monomers, volatile	Gas chromatography (GC) and head-space techniques
from 150 to 600	Volatile	
from 200 to 1,000		NMR, liquid chromatography
from 40 to 1,000	Presence of a chromophore	Direct UV spectrophotometry

Table 8.1. *Analytical methods used for analyzing potential migrants*

8.3.2.2. *Types of migration*

Potential interactions between food and packaging have been the subject of studies where migration occurs from the packaging to the food. However, food constituents can also migrate to the packaging material: this reverse phenomenon occurs in particular for most plastics, which can affect the mobility of certain additives in polymers. Materials can be classified into three groups.

Class 1: No or negligible migration

A range of product-packaging combinations exist for which virtually no migration is observed. This is particularly the case in dry conditions where there is practically no exchange between the food and most packaging materials.

Class 2: Migration independent of the packaged product

This class mainly includes plastics: polymers contain a certain amount of monomers as reaction components mainly originating from the reaction process. Monomers and other products can also be formed from polymers by thermal degradation during the manufacture of packaging or photo-degradation by radiation (ultraviolet or ionizing). These are small molecules (ethylene, propylene, styrene, vinyl chloride) that can diffuse through the polymer by a concentration gradient, even in the absence of external forces (vibrations).

The contamination of food products by the diffusion of gaseous monomers is less critical in so far as they are volatile and as long as they are below the legal limits at the time of package opening.

The amount of product that migrates by diffusion follows Fick's law; it is proportional to time, the surface of the packaging and the initial concentration of the migrant in the packaging material. The thickness of the packaging is almost irrelevant once it exceeds 15 µm. After a certain amount of time (around 5 days at 40°C), an equilibrium is reached between the migrant in the material and the surrounding medium and therefore migration slows down or stops.

Class 3: Migration linked to interaction with the product

In this case, migration is controlled by the properties of the two phases that are in direct contact, such as with liquids. Three possibilities can arise if the packaging material comes into contact with the contents.

– *Chemical or electrochemical reactions* between the packaging wall and the contents (with metal packaging): metals (tinfoil, aluminum) can react with acids in the food (especially fruit and vegetables) to form hydrogen gas and metal ions causing corrosion. In the case of cations, the general reaction can be as follows:

$$M + 2H^{+} \rightarrow H_2 + M^{2+} \qquad [8.9]$$

Tin, which forms the outer layer of cans, is covered with a thin layer of oxides, which reduces its reactivity to acids. However, this layer is not resistant to highly acidic media, which is why products with a pH below 4–4.5 are usually packaged in cans covered with epoxy-based protective coatings.

Various electrochemical phenomena can occur in which the metal substrate acts as an anode or cathode; the ionic species formed can contribute to the deterioration of sensory quality (color and flavor). This phenomenon occurs where the two metals in a can (e.g. iron and tin) come into contact with the liquid at the pores of the coating. According to legislation, damaged cans cannot be sold since various elements in the packaging material may come into contact with the product. Other reactions are possible between the metal packaging and the contents such as the reaction between tin and the sulfur in tomato puree or peas for example. Tin sulfide is formed, which is insoluble, but highly colorful, dotting the metal can with black or brown spots. Although this does not contaminate the product, it is visually less appealing to the consumer.

In conclusion, the reactivity of metals largely depends on the purity and quality of tin deposition and surface treatment, as the presence of impurities results in greater reactivity.

– *Ion exchange between the packaging material and the food* (glass or ceramic packaging). The glassy state is a supercooled liquid of silicon dioxide, which is the main component of glass. Other elements such as sodium, calcium, magnesium, cadmium and lead (common metals used in the production of crystal glass and glazes) are present for practical and economic reasons. Different types of silicates are embedded in the base structure formed by silicon dioxide. Thus, the surface of glass can act as an ion exchanger when it comes into contact with solutions containing ionic species: proton and cation exchanges (sodium, cadmium, etc.) can occur in the presence of acid. For most glass and ceramics used, these exchanges are limited given the low concentrations. For example, a one-liter bottle (product/packaging contact surface of 500 cm^2) transfers 25 mg of sodium in six months. Nevertheless, limits are set by law on the migration of lead and cadmium, given their toxicity.

– *Absorption of liquids* by the packaging material (mostly plastic or paper packaging). Plastics can contain two groups of non-polymeric products: intentional additives to modify the physical, chemical and mechanical properties of plastics in order to facilitate their manufacture, use and recycling (antioxidants, plasticizers, anti-blocking agents, anti-mist agents, stabilizers, etc.), and unintentional additives such as the residues from polymer synthesis (monomers, catalysts, solvents), impurities from raw materials, substances derived from polymers and additives formed during the manufacture or use of packaging.

Plastics are susceptible to migration, the extent of which depends on the chemical nature of the liquid food product (water, alcohol, acid and oil). Liquid products tend to penetrate into the plastic material causing it to expand; additives can then leak from the liquid medium absorbed in the packaging into the product. The migration of additives is therefore controlled by the rate of penetration of food into the packaging material. For paper packaging, the extraction of migrants is very rapid as the penetration of liquid into the packaging is very fast.

8.3.2.3. *Migration tests and legislation*

Legislation requires that packaging materials be inert to the food in contact, that is materials should not transfer their constituents to foodstuffs in quantities which could endanger human health or bring about an unacceptable change to the composition or sensory quality of the food. Migration tests are conducted for control purposes; there are specific migration tests designed to measure the transfer of identified constituents and overall migration tests designed to measure all constituents that have diffused from the packaging to the food. In most cases, when overall migration is measured, neither the number nor the type of migrating constituents is known.

Given the fact that EU legislation is constantly changing in this area, it is impossible to provide a general protocol that applies to all packaging materials as each class of material requires specific methods. For example, migration tests exist for plastic packaging. Migration is measured using different model solutions (distilled water, 3% (w/v) acetic acid in aqueous solution, 15% (w/v) ethanol in aqueous solution, rectified olive oil) depending on the food to be packaged (Table 8.2).

	Water	*3% acetic acid*	*15% ethanol*	*Olive oil*
Non-alcoholic beverages	X	X	-	-
Pasta	-	-	-	-
Chocolate	-	-	-	X / 5
Oils and fats	-	-	-	X
Fish	X	-	-	X / 3
Vinegar	-	X	–	–
Processed cheese	X	X	-	-
Chips	-	-	-	X / 5

The X indicates the model to be used. The result of the migration test should be divided by the number after the X; this is known as the reduction coefficient and takes into account the higher extraction capacity of model fatty foods compared to other types of fatty foods.

Table 8.2. *Examples of liquid models for different foods according to the directive on the conventional classification of food (according to [LOX 98])*

Migration test conditions (time, temperature) must correspond to those used during packaging and storage (Table 8.3). If the packaging can be used under any time and temperature conditions, tests lasting 10 days at 40°C or two hours at 70°C are carried out, which cover the most extreme conditions.

If these tests result in physical or other changes to the packaging, more suitable conditions should be chosen.

Normal conditions of contact	Test conditions
Contact time greater than 24 h	
$\theta \leq 5°C$	10 days at 5°C
$5°C < \theta \leq 20°C$	10 days at 20°C
$5°C < \theta \leq 40°C$	10 days at 40°C
Contact time between 2 and 24 h	
$\theta \leq 5°C$	24 hours at 5°C
$5°C < \theta \leq 40°C$	24 hours at 40°C
$40°C < \theta$	in accordance with national legislation
Contact time less than 2 h	
$\theta \leq 5°C$	2 hours at 5°C
$5°C < \theta \leq 40°C$	2 hours at 40°C
$40°C < \theta \leq 70°C$	2 hours at 70°C
$70°C < \theta \leq 100°C$	1 hour at 100°C
$100°C < \theta \leq 121°C$	30 minutes at 121°C
$121°C < \theta$	in accordance with national legislation

Table 8.3. *Comparison between the conditions of use of plastics and test conditions (according to [FEI 98])*

EU directives provide a list of raw materials and additives that can be used in the manufacture of packaging, specifying the limits for specific and overall migration, and define the maximum permitted quantities of residual substances and impurities from the manufacture and processing of materials. This legislation covers packaging as well as all materials intended to come into contact with foodstuffs. According to Directive 90/128/EEC, the limit of overall migration of permitted substances in contact with food, with the exception of those mentioned in the list, is 10 mg dm^{-2} or 60 mg kg^{-1} in the following cases: containers or articles comparable to containers with a capacity of 500 ml to 10 l, containers or articles comparable to containers where it is not possible to estimate the surface area in contact with foods, and sealing devices (caps, gaskets, stoppers, etc.). The specific migration limits are unique to each component [SCH 07].

8.3.3. *Other properties of packaging*

8.3.3.1. *Mechanical strength*

Packaging protects products from external impact by its mechanical strength, which depends on the rigidity of the packaging material. Products can also be protected by gas packaging (e.g. bags of crisps). The strength of the material can be evaluated by burst, compression and perforation tests. Packaging must also be able to withstand changes in internal pressure, as in the case of sterilized products. Strict standards (European directive GAMMES, 1980) are set for pressurized containers (aerosols).

8.3.3.2. *Thermal conductivity*

Insulating materials may be used where variations in positive or negative temperatures could affect product shelf life and quality. Conversely, it is preferable to choose packaging with high thermal conductivity where heat treatment (sterilization) is carried out, which explains the use of metal cans for sterilized products. If, for reasons of mechanical strength, metals are used in packaging, it is important to be aware of possible thermal bridges that could easily negate the insulation properties of the container. The insulating power of some materials can be improved by trapping a gas (e.g. air), as in the case of low density materials (foams, expanded polystyrene, wood, etc.).

8.3.3.3. *Radiation barrier properties*

It is generally ultraviolet rather than visible light rays that tend to denature products. Products can be preserved by opaque packaging or, where the packaging should remain transparent (e.g. beverages), tinted packaging. It is also possible to use an outer layer of cardboard or wood.

Ionizing radiation is permitted to disinfect certain packaging materials, including plastics. However, studies have shown that volatile compounds (hydrocarbons, ketones, aromatic compounds) are formed during electron beam irradiation of most food-grade plastic packaging (polyethylene, polyester, oriented polypropylene). The amount of compounds increases with the dose of ionizing radiation applied.

8.4. Packaging materials

8.4.1. *Cellulosic materials*

Traditionally, packaging materials were of plant origin (leaves to carry berries, etc.). It is therefore not surprising that cellulose fibers are still used today as packaging materials (Figure 8.3).

8.4.1.1. *Wood*

Raw wood is a natural, homogenous material, composed of strong fibers connected by an elastic plastic structure; it has many suitable properties for packaging: lightweight, good mechanical and chemical resistance, thermal insulation, good hygroscopicity (absorbs water without condensation), easy to assemble and aesthetic. These properties can be modified to suit the different types of packaging: plywood (cross-grained strips of wood), chipboard (chippings glued together), wood veneer (thin slices of wood obtained by peeling the trunk of a tree, often used for cheese boxes, crates and baskets), or laminated timber. The main types of wood suitable for food are pine, oak, poplar and cork oak for the production of corks. Raw wood is used in the packaging of consumer food products, whether at primary (trays for fruit and vegetables), secondary (cheese boxes, wine crates) or tertiary level (palettes) [COU 98].

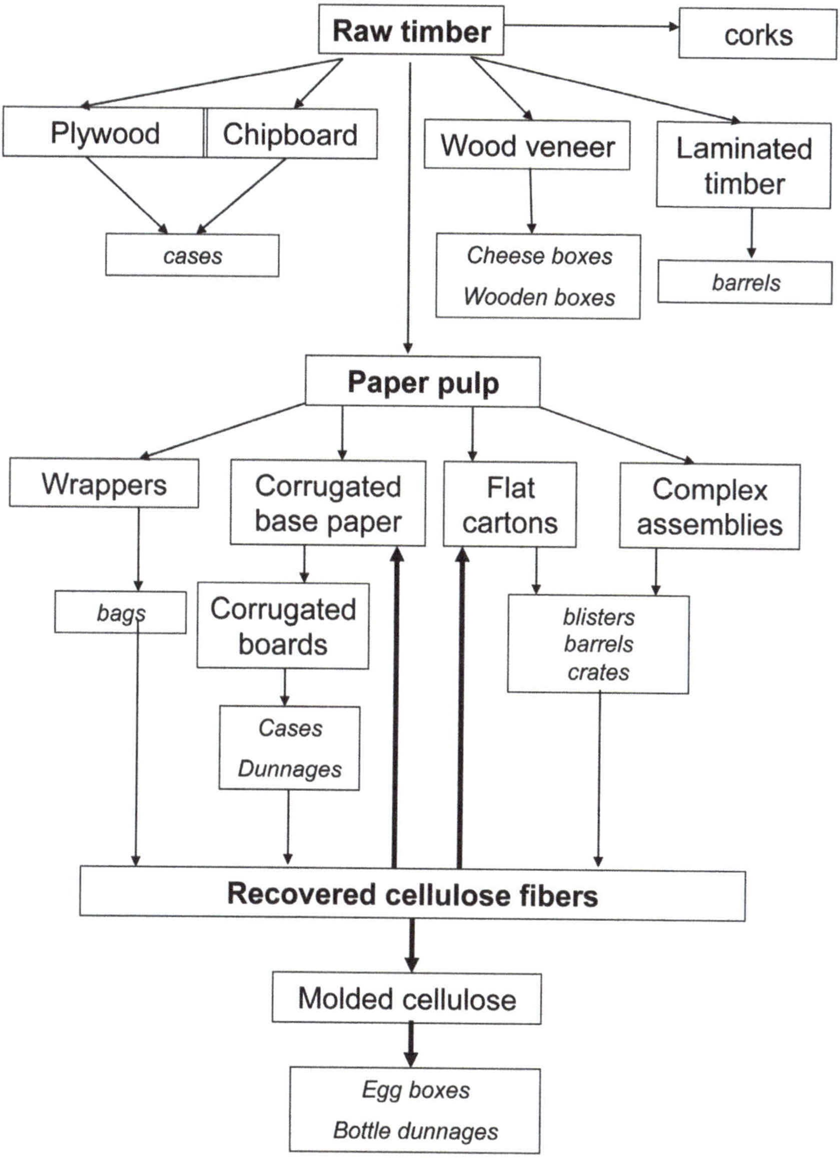

Figure 8.3. *Cellulosic materials used in packaging*

8.4.1.2. *Paper and paperboard*

Paper is produced by pressing thin fibers into thin sheets. Paper pulp is mainly produced from soft wood (birch, eucalyptus, poplar, conifers) to which recovery or recycling products are often added (Figure 8.4). There are two methods for producing paper pulp:

– a mechanical process in which the wood is ground and mixed with water, forming a brown slurry containing all the wood components including fibers of varying lengths, but also tannins, resins, etc. This method is used to make boxboard;

– a chemical process where grinding is followed by the chemical separation of fibers from other components using soda, sulfite or sulfate. The soda process is used to obtain paper for the manufacture of corrugated fiber board as the pulp remains colored. The sulfate process partially attacks cellulose, which produces kraft paper after short treatment and highly resistant paper after long treatment.

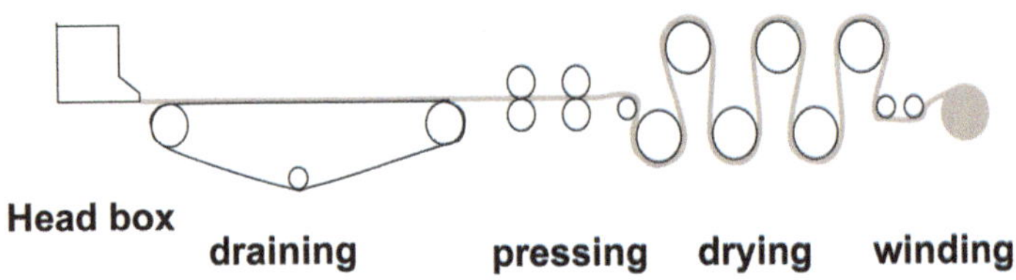

Figure 8.4. *Manufacture of paper*

The pulp is refined during which various additives are added to obtain the final paper quality; these include bonding agents (hydrolyzed starch, alum or methylcellulose) to facilitate printing, as well as mineral fillers (kaolin, talc, calcium salts, titanium dioxide) to improve whiteness and remove surface irregularities. Finally, polyacrylamide, urea-formaldehyde, synthetic latex or polyethylenimine are added for greater mechanical strength, especially when the paper is wet.

Paper is an inexpensive, renewable and recyclable product, making packaging more rigid but having no barrier or wettability properties. It is therefore often used in combination with aluminum foil and a polymer for sealing purposes. Paper coated with polyvinylidene chloride (PVDC) also exists, which acts as an effective barrier to water and oxygen (Table 8.4).

Term	*Pre-processing*	*Properties and applications*
Glassine	Aluminum foil lamination	*Butter, cheese, ice cream*
Kraft Paper	Aluminum foil lamination Polyethylene extrusion PVDC coating Hot melt adhesion	Whiteness and excellent printability e.g. *soups, chocolate, biscuits, sealing*
Calendered Kraft Paper	Same process as Kraft Paper plus waxing	Acceptable printability cheese, confectionery
Genuine Parchment	Lamination Polyethylene extrusion Hot melt adhesion	Does not disintegrate, high moisture resistance, good resistance to fat
Bleached Greaseproof Paper		Greaseproof for pet food e.g. *coffee liner bag, labeling of fat products, biscuit packaging*
Glazed Bleached Kraft	Aluminum foil lamination Polyethylene extrusion	High mechanical strength Good printability

Table 8.4. *Overview of paper grades and their applications*

The difference between paper and paperboard is its weight. The term paperboard applies beyond a weight of 250 g.m^{-2} [RIL 12a]. There are two types:

– *corrugated fiberboard* is obtained by gluing a sheet of corrugated fluted paper between two sheets of liner paper (Figure 8.5). This structure has good mechanical strength and thermal insulation (trapped air) without any significant increase in weight. However, it is not suitable for printing, which means that it may be necessary to print before the flutings (corrugated layers) are glued to the liners. The advent of nanotechnology (nano-corrugated board) has improved the quality of printing.

– *boxboard* or *cartonboard* is made up of multiple layers of cellulose fiber. Its main quality is that it can be decorated, that is cut, embossed, scored (to facilitate folding), pre-cut (for easy opening), coated (to improve its barrier properties), rendered impermeable to oil or grease (greaseproof), anti-fungal treated, etc. Different types of packaging are made from boxboard: cases, tubes, boxes, packing dividers between bottles, etc.

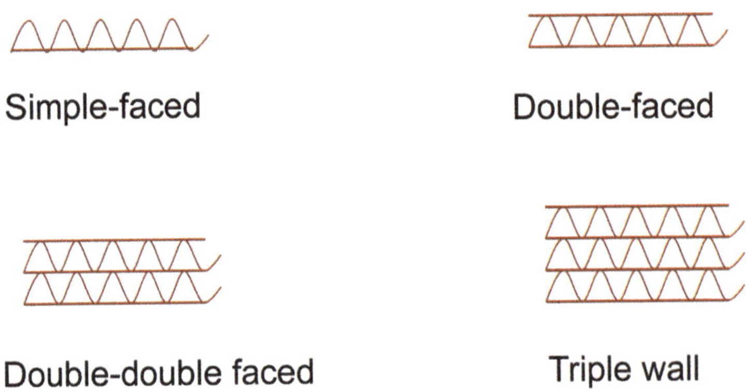

Figure 8.5. *Different categories of corrugated paperboard*

8.4.1.3. *Molded pulp*

Molded pulp is made by shaping recycled paper pulp in mesh moulds. It is used to produce egg cartons and inner packaging.

Cellulosic material has the advantage of being biodegradable and recyclable at a lower cost than virgin fibers derived from wood. However, its mechanical properties are inferior to those of virgin fibers due to the gradual decrease in fiber length during recycling. In addition, its level of whiteness is often insufficient. Finally, its use in the production of paper requires prior decontamination (various metals, plastics, inks, etc.).

8.4.2. *Glass*

Glass is one of the oldest man-made packing materials. It is a mineral product obtained by melting, which solidifies without crystallizing. It is composed of 70% silica (glass-forming agent), 14% sodium carbonate (fluxing or melting agent), 10% lime, 1% magnesium, 1% potash (stabilizing agents), metal oxides for color and filtering properties [GRA 12].

Glass manufacturing is carried out in an integrated and continuous process, which means that the final product can be obtained in a single plant, which is not the case for other types of packaging (metal, plastic and cardboard). Raw materials are poured into a melting furnace at a temperature of 1,550°C. Glass is then thermally conditioned and circulated to machines via a working tank and distribution channels that deliver gobs (drops of molten glass) to the forming machine. It then undergoes surface treatment to prevent scratches and other abrasions that might weaken its mechanical properties, and finally thermal annealing treatment at 550°C to remove any stresses generated during forming in the machine. The product is then subjected to quality control before being used.

Glass packaging typically includes bottles, jars, flasks, glasses and tumblers. It is used in many different sectors of the food industry (beverages, preserves, jams, condiments, baby food, dairy products, etc.). A distinction is made between glass varieties depending on their ability to absorb heat radiation and block ultraviolet light: clear glass for water, certain juices, jams and yoghurts; green/blue glass for beer, wine and oil; brown/amber glass for beer and certain juices.

Glass has many qualities: impermeable to gas, vapors and liquids (excellent barrier), good chemical inertia (higher than that of all other packaging materials), easy to clean and sterilize, odorless and does not transmit flavors, transparent, colorable (protection again UV light for example), rigid, resistant to high internal pressures (fizzy drinks), transparent to microwaves, inexpensive and recyclable. Its main drawback is its risk of breaking. In addition, proper closing of glass packaging is also essential to ensure tightness of the container and consequently protection of the product. A wide range of closures exist (cork, metal, plastic); they must ensure hermetic sealing of the packaging and proof of no tampering upon first opening, and allow easy opening as well as resealing where the product is only partially consumed.

8.4.3. *Metals*

Metals are widely used for food packaging given their suitable properties: shaping ability, rigidity, solidity, impermeability, opacity with regard to light rays, heat conduction, etc. They are mainly used in the packaging of canned goods since they are particularly well suited for long-term storage (robustness and impermeability). Moreover, metal packaging is generally recyclable. Two metals are used in packaging: steel in the form of tinplate (tin coated steel) or chromium coated steel and aluminum in the form of alloys.

8.4.3.1. *Steel*

Steel is an alloy of iron ore (magnetite) and carbon (coke). Both products are heated to a high temperature in a blast furnace to reduce the iron ore (Fe_3O_4) to iron (Fe). The liquid iron is then continuously cast into slabs, which are then hot or cold rolled into 0.2 mm sheet metal. The next process is electrolytic tin plating, whereby the deposit is melted again to obtain an alloy with the support. The surface is then treated to prevent thermal collapse and protected using zinc, chromium or nickel. Varnishing can also be carried out to limit corrosion and migration upon contact with the food product. Sheet metal is used in steel packaging in the production of two- and three-piece cans. Two-piece cans are obtained by reforming a disc of metal into a cylinder with an integral end, to which a loose end is seamed to close the can. The operation of reforming sheet metal without changing its thickness is called "drawing". The operation of thinning the walls of a two-piece can by passing through circular dies is called "ironing". Three-piece cans are obtained by stapling, seaming or gluing a molded cylindrical body to pressed can ends (lids). Sheet metal is also used to make trays, drums, aerosol containers, etc. [PAG 12].

8.4.3.2. *Aluminum*

Aluminum can be extracted from aluminum ore by grinding and mixing with soda at high temperature and pressure. The resulting aluminum oxide, also called alumina, is the raw material in the production of aluminum by electrolysis. Alumina is added to tanks through which a current of more than 300 kA passes, and is mixed with molten ore (cryolite) to facilitate the extraction of aluminum metal. Under the effect of the current, oxygen is formed at the anode, while aluminum forms at the cathode and sinks to the bottom of the tank. The molten metal is regularly removed and sent to a foundry where it is refined and turned into an alloy depending on the end product. It is then cast into slabs, billets, ingots or wire.

Several aluminum alloys are used in packaging; varying percentages of magnesium, manganese or chromium give them different properties (mechanical resistance, susceptibility to corrosion, shaping ability). Aluminum packaging includes cans (only two-piece), trays and foil. Its malleable nature makes it easy to open or tear depending on the packaging product [PAG 12].

8.4.3.3. *Protective coating of metal packaging*

The main drawback of using metals as packaging material is their susceptibility to corrosion, which is well controlled in most cases by choosing the most suitable material for the specific product. However, due to the wide range of food products (composition, filling conditions and storage), corrosion cannot be completely avoided. As a result, metal is often coated with a protective layer, similar to plastics.

The organic coatings of metal packaging have the physicochemical properties of polymers (insolubility and chemical inertness in aqueous media). They have excellent mechanical properties (adhesion, hardness, flexibility); adhesion and absence of porosity are the two main factors that determine the protective quality of these coatings. They range in thickness from 5 to 10 µm.

Group	*Application*	*Flexibility*	*Adhesion*	*Resistance to sterilization*
Oleoresins	Fruit and vegetables (anti-sulphur coating)	Poor	Good	Average
Phenolics	Fruit, vegetables, meat (barrier coating)	Poor	Poor	Very good
Epoxy phenolics	Wide range (fruit, vegetables, meat, etc.)	Good (depends on epoxy/phenolic ratio)	Good	Good
Vinyls	Beverages (beer, carbonated drinks)	Excellent	Good	Good
Organosols	Wide range for drawn cans	Very good	Very good	Good
Acrylics		Good	Very good	Average
Epoxy-urea	Beverages	Good	Good	Average
Polyesters		Average	Good	Good

Table 8.5. *Main groups of organic coatings*

Several types of coatings are used depending on the kind of food (Table 8.5) and the type of can used. Thus, for preserves in three-piece cans, the coating should be without defects and resistant to sterilization. With the development of drawn cans (two-piece cans), mechanical deformation is greater, and adhesion and flexibility should also be taken into account. There are generally three categories of products based on type and container-content reactions:

– *high-sulfur foods*: proteins can release sulfur compounds during the sterilization process. They readily react with iron and tin to give metal sulfides, which are brown or black compounds that neither alter the flavor nor the nutritional value of the food but which significantly impair the appearance. To overcome this, cross-linked coating is used (phenolic or epoxyphenolic resin), to which pigments can also be added;

– *acidic foods*: do not necessarily require the presence of a coating, but tin is sometimes added as it helps remove oxygen that could oxidize the product. Its reducing role is important in clear fruit juices. In cases where a coating is recommended to prevent corrosion of the container, organosols with or without an epoxy phenolic resin are usually used;

– *beer*: due to its extreme sensitivity to metal contamination and *carbonated drinks*, some of which are highly corrosive, requires maximum protection. Bi-layer systems are used (epoxy phenolic/vinyl, epoxy-urea/vinyl, epoxy-urea/epoxy-urea).

Thus, the choice of internal organic coatings depends on the reactivity of the food, the type of can (drawn (two-piece) or assembled (three-piece)), the type of lid (with or without easy opening) and the intended shelf life of the canned product [APP 98].

8.4.4. *Plastics*

Plastics are synthetic materials composed primarily of macromolecules and can be molded under heat and pressure. In chemical terms, a plastic comprises a macromolecular organic phase (polymer or resin), fillers or reinforcing agents (glass, fibers, etc.) and additives (plasticizers, heat stabilizers, anti-UV agents, colorants, etc.). In general, there are two types of resin: thermoplastics, which soften when heated and harden when cooled and can be constantly remolded, and thermosets, which can only be shaped once. Plastics are mainly derived from the petrochemical industry except for cellophane, which is obtained by the chemical treatment of cellulose.

8.4.4.1. *Composition*

Polymers

Polymers are long chains of monomers such as polyvinyl chloride (PVC) consisting of vinyl chloride monomers (CH_2=CHCl) or polypropylene (PP) consisting of propylene monomers. Polymer structures can differ: linear structures with one type of monomer (homopolymer) or a combination of two types of monomers (copolymer), branched structures of homopolymers or graft copolymers that tend to reduce molecular mobility, or three-dimensional polymers. Polymer properties are determined by their three-dimensional arrangement (coexistence of amorphous and crystalline regions) and the chain density (e.g. high and low density polyethylene).

All plastics offer good barrier and safety properties even in monolayer packaging: they are referred to as structural materials. Barrier polymers are used for foods that are highly susceptible to oxidation, light or flavor loss, which are used in multilayer packaging together with structural materials (Figure 8.6):

– *structural polymers* (Table 8.6) are chosen depending on the type of packaging required for the product; they offer a wide variety of properties such as resistance to freezing, sterilization and microwaves, attractive appearance, user-friendliness, etc.;

– *barrier polymers* (Table 8.7) have very low permeability to oxygen, carbon dioxide and flavors. They are currently the preferred material, but are never used in monolayer packaging due to cost limitations.

Copolymers such as ethylene vinyl alcohol (EVOH) are also commonly used; they combine two polymers and exhibit additional properties. Copolymers should not be confused with multilayer systems whereby extra properties can also be obtained.

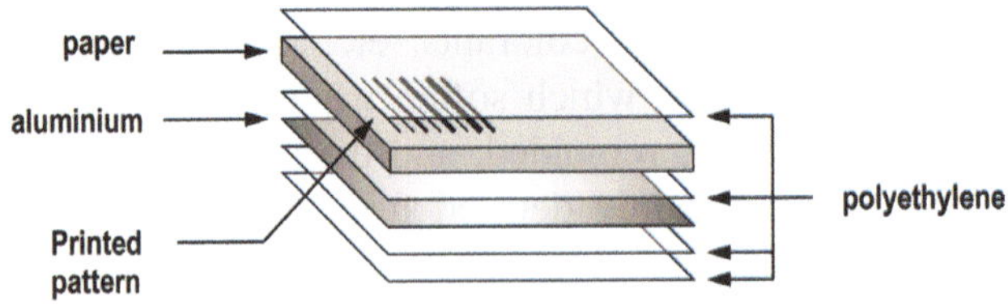

Figure 8.6. *Multilayer packaging*

Polymer	Properties					
	Permeability		*Resistance*	*Susceptibility*	*Others*	*Applications*
	Water vapor[a]	O_2[b]				
Low density polyethylene (LDPE)	10	7,000	UV, non-oxidizing acids and bases, polar solvents		Sealable, flexible, printable, transparent	Sacks, bags, boxes, tubes, bottles, shrink film, stretch film, trays, seals
High density polyethylene (HDPE)	5	2,100	Mechanical shock, sterilization		Rigid, opaque, printable	Bottles, tubes, caps
Polypropylene (PP)	6 to 10	1,800 to 3,600	Mechanical and thermal stress (microwave, sterilization), mineral aqueous solutions, dilute acids and alkalis	UV, oxidation	Rigid	Films, boxes, trays (ready meals), blister trays (biscuits), tubs (yoghurt), stoppers, tubes
Polyvinyl chloride (PVC)	35	120	Chemical products, fats	Heat, light and oxidation	Rigid, glossy, transparent	Films, bottles (oil, vinegar, wine, syrup), trays, boxes

Polystyrene (PS)	140	4,000	Dilute acids, aqueous and alcohol solutions	UV, oxidation and organic solvents	Rigid; transparent (crystal polystyrene GPPS); opaque, glossy, breakable (anti-shock polystyrene HIPS); light, thermoformable	Boxes (eggs), tubs (yoghurt), disposable cups, caps, wrapped trays (meat)
Expanded polystyrene (EPS)					Isothermal, moldable	Trays (meat and fish)
Polyethylene terephthalate (PET)	32	65	Mechanical shock, esters, aromatics, alcohols, fats, dilute acids and alkalis, oxidation, light		Glossy, printable, transparent	Blowmolded containers (replaces PVC) for liquids (alcohol, water)
Polyamides (PA)	50 to 170	3 to 50	Chemical products	Acids, amines, light	Flexible	Wide range (more than 30 PA currently used)

[a]) in [g m^{-2} day^{-1}] at 38°C, 90% H_R, 25 µm thickness
[b]) in [$cm^3 m^{-2}$ $day^{-1} atm^{-1}$] at 23°C, 0% H_R, 25 µm thickness

Table 8.6. *Main structural polymers: properties and applications (according to [DEL 98])*

Polymer	Properties				Combinations
	Permeability		Susceptibility	Others	
	Water vapor[a]	O_2[b]			
Ethylene vinyl alcohol (EVOH)	22 to 60	0.1 to 1.4	Moisture	Impermeable to CO_2 and flavors	With water-impermeable polyolefins (PE or PP)
Polyvinylidene chloride (PVDC)	0.5 to 3	0.6 to 10	Heat	Impermeable to CO_2 and flavors	With PS, PVC, PC
Polyacrylonitrile (PAN)	95	12		Impermeable to CO_2 and flavors	

[a]: in [g m^{-2} day^{-1}] at 38°C, 90% H_R, 25 µm thickness
[b]: in [cm^{3}m^{-2}day^{-1}atm^{-1}] at 23°C, 0% H_R, 25 µm thickness

Table 8.7. *Main barrier polymers: properties and combinations (according to [DEL 98])*

Fillers

Fillers are inert organic or inorganic compounds, whose role it is to modify mechanical, thermal and electrical properties, improve surface quality and reduce the cost of plastics. These compounds are dispersed in the polymer matrix. There are spherical fillers (powder or flour) that improve the flowability of resins and their resistance to compression, fibrous fillers (cellulose, glass) that increase tensile strength and rigidity, and mineral fillers (chalk, silica) that improve electrical properties, resistance to heat and moisture and increase density.

Additives

Additives are generally organic or organometallic compounds of lower molecular weight than polymers that are capable of modifying the physical or chemical properties of the polymer. They include:

– *plasticizers* to improve the mechanical properties of plastics by insertion between the molecular chains;

– *stabilizers* to prevent deterioration of the material during manufacture or use: antioxidants for polyolefins (PP, PE, PS) that are particularly susceptible to oxidation, anti-UV for PET, PS and PA that absorb ultraviolet light and can thus photodegrade or photooxidize, antiozonants for rubber that is highly susceptible to ozone, fungicides, etc.;

– *lubricants* to facilitate workability in calendars, extruders and presses, and reduce resin/metal friction;

– *antistatic agents* to reduce or eliminate the build-up of static electricity by making the surface slightly conductive and therefore dust-repellent;

– *flame retardants* that inhibit or resist the spread of fire;

– *colorants* that help meet the light stability requirements of the product, etc. The current practice is to use organic instead of inorganic pigments containing heavy metals, despite the fact that they offer lower opacity and less resistance to migration and UV.

Additives are used for their primary function, but may have side effects that can be controlled by the addition of another additive. If they reach their solubility limit in the polymer, they can exude to the surface. They are carefully monitored by law because their low molecular weight allows them to migrate easily.

8.4.4.2. *Processing techniques*

Several different processes exist for the manufacture of plastics [RIL 12b]:

– *extrusion* involves melting and mixing using a rotary screw, during which polymerization takes place. It is a continuous process and is thus used to manufacture long segments. The polymer is shaped once it exits the die;

– *extrusion blow molding*; in this process, plastic is melted and extruded into a hollow tube (parison). Air is then blown into the parison, inflating it into the shape of the mold, resulting indifferent mechanical strength and barrier properties; this is done for certain plastic bags for example;

– *injection molding* consists in injecting plastic into a mould. This process ensures the uniform distribution of the material;

– *thermoforming* involves stretching a sheet of plastic by suction, blowing or stamping onto the wall of the mould. Unlike injection, it is difficult to evenly distribute the material, and the edges are often thicker. This technique is used to produce multilayer packaging, which is not possible with injection;

– *injection stretch blow molding* involves forming packaging from injected or thermoformed preforms, which have been manufactured in a previous stage. This technique reduces the transport of packaging, but requires that the manufacturer is equipped with a stretch blow molding machine.

8.4.5. *Biomaterials*

Increased awareness among consumers with regard to environmental issues (especially with the introduction of waste separation and recycling) has encouraged the development of new recyclable and biodegradable packaging materials. They can be divided into four main categories of biopolymers: polysaccharides, proteins, lipids and polyesters (obtained from plant or bacterial biosynthesis). Films consisting of polysaccharides (cellulose and derivatives, starch and derivatives, gums, etc.) or proteins (gelatine, zein, gluten, etc.) generally have good mechanical and optical properties, but are very sensitive to moisture and have poor water vapor barrier properties. However, films made of lipids (waxes, lipids and derivatives) or polyesters (polylactic acid (PLA)) have good water vapor barrier properties but are usually opaque and difficult to shape. They are also very fragile and not very stable (susceptible to rancidity) [GON 94].

8.4.5.1. *Use of biopolymers in the production of packaging*

Three different techniques using renewable agricultural raw materials have been developed for the production of bio-packaging:

– mixtures of synthetic polymers and biopolymers, which constitute oxo-biodegradable rather than biodegradable packaging, with the aim being to make synthetic polymers more susceptible to microbial attack. The most commonly used material is starch; other combinations include cellulose/polyurethane, gluten/synthetic resin, plant protein/vinyl compound, and casein/synthetic polymer mixtures.

– the use of agricultural raw materials as fermentation substrates for the production of recyclable and biodegradable polyesters and bacterial polymers. This category includes polylactic and polyglycolic acids, polycaprolactone and chitosan. The limiting factor of these biomaterials is their extraction and purification cost;

– the direct use of agricultural polymers generates less sophisticated low-cost packaging given the reduced cost of raw materials; they are fully biodegradable and edible if no non-food additives have been added. The most used raw material is starch, since it is inexpensive, widely available and relatively easy to handle.

8.4.5.2. *Properties and applications of bio-packaging*

Bio-packaging must meet the same requirements as packaging in terms of barrier, mechanical, non-toxicity properties, etc. These properties depend on the type of material used, its modc of production and its applications. Plasticizers, cross-linking agents, antimicrobials, antioxidants or texturizing agents can be added to improve the functional properties of bio-packaging, but may increase its environmental impact. Materials based on biopolymers have high gas barrier properties, especially with respect to oxygen. Lipid compounds also have good water barrier properties. The main applications are the production of bags, trays, inner packaging and inserts.

One of the applications of bio-packaging is edible packaging, that is films and coatings for certain products (wax coating on fruit, chocolate or sugar coating for confectionery, etc.); coatings are formed directly on the food whereas films are formed separately and then applied to the food. These edible films and coatings must have sensory properties that are as neutral as possible. They can be used to improve the surface appearance or feel of the product, encapsulate colorings, flavorings and spices to maintain a high concentration, etc.

8.5. Packaging technologies

8.5.1. *Vacuum packaging*

Vacuum packaging first appeared in the 1960s. The objective is to remove the main spoilage agent from the food environment, which is oxygen. Vacuum packaging achieves a residual oxygen content of approximately 1%. This can be further reduced by tissue respiration of the product (in the case of meat for example) or bacteria, and replaced by carbon dioxide. The combination of these two effects (depletion of oxygen and accumulation of carbon dioxide) is responsible for the inhibition of aerobic spoilage flora and oxidation [GUE 98].

This packaging method requires a high quality product, especially from a microbial perspective, suitable packaging that ensures mechanical strength, sealability and maximum impermeability to different gases as well as the appropriate packing machines. Two types of packaging are usually used:

– *semi-rigid packaging* consisting of a thermoformed plastic tray covered in film. The machine is supplied with two rolls of film: an inner film thermoformed into a tray into which the product is placed and an outer film that serves as a lid. Vacuum is created in the tray before the covering film is hermetically sealed. This type of packaging is often used for meat products;

– *bag packaging* using gas flushing machines. In these machines, air is pushed out of the packaging by continuous flushing with another gas.

In the case of flexible or rigid, large and medium packaging, the compensated vacuum technique can be used by means of a two-way valve.

Vacuum packaging allows food, especially prepared meals, to be cooked. Vacuum cooking is meant in the strict sense where the raw product is vacuum packed and then cooked at a low temperature ($\theta \leq 100°C$) and rapidly cooled. This technique was made possible by the introduction of heat-resistant flexible plastics, which remain chemically inert during cooking and storage, are impermeable to gases and aromas and maintain mechanical strength at high temperatures. Most of these materials are multilayer; they include non-shrink bags often made of a polyamide base and a polypropylene bonding agent, used for example with sauce dishes. Due to their relative permeability to gas, they cannot be stored for long periods, but this is not necessary for the type of products involved. There are also coextruded shrink bags with high barrier properties intended for cooking meat, whole fish, etc. Shrinkability is a property that reduces exudates during cooking.

There are, however, some disadvantages to vacuum packaging such as the fact that it promotes conditions for anaerobic bacterial growth or, in the case of red meat, causes discoloration. In addition, it is difficult to obtain and maintain proper vacuum in a package. As a result, this type of packaging has gradually been replaced by modified atmosphere packaging.

8.5.2. *Modified atmosphere packaging*

This type of packaging first appeared in France in the mid-1970s. Its objective is to extend modified atmosphere packaging shelf life, preserve sensory properties and present the product in a more attractive manner by limiting physical, enzymatic, biochemical and microbial degradation. The machines used are identical to those used for vacuum packaging and the gas mixture is injected into the packaging after vacuum treatment [SIV 02].

8.5.2.1. *Function of gases*

The three main gases used in modified atmosphere packaging are nitrogen, carbon dioxide and oxygen. It is also possible to use helium, argon and nitrous oxide (Table 8.8). The proportion of gas in the package should be around one third of the volume with the product taking up the rest.

Nitrogen

Nitrogen is primarily used to replace oxygen in the package to reduce the oxidation of pigments, aromas and fats. It is an inert, odorless gas and is poorly soluble in water or fats, thus preventing shrinkage in the package. It is also used to prevent crushing (e.g. crisp packets).

Carbon dioxide

Under certain conditions (more than 20% in the atmosphere at equilibrium), carbon dioxide is a bacteriostatic and fungistatic agent that can delay the exponential growth phase and reduce the proliferation rate of moulds and aerobic bacteria. High concentration CO_2 has a selective inhibitory effect, to which moulds and bacteria from the *Pseudomonas* and *Achromobacter* genera are very sensitive. By comparison, yeasts and lactobacilli are more resistant, even insensitive. The inhibitory effect of CO_2 is more pronounced if packaging is carried out immediately after production when microorganisms are still in the latent phase and the microbial load is lower. Furthermore, the inhibitory effect of CO_2 increases as the temperature drops due to its improved solubility in the aqueous phase of the product. It is therefore highly recommended to refrigerate CO_2 modified atmosphere packaged products.

The mode of action of CO_2 is still not entirely known and several hypotheses have been proposed, for example, reduction in intracellular pH resulting in a slowdown of enzymatic activity, specific inhibition of decarboxylation enzymes or non-specific inhibition of other enzymes, and modification of cell membrane properties.

Due to its solubility in water and fats, it can sometimes cause shrinkage of the packaging film.

Oxygen

Oxygen is usually the gas that is being replaced. However, it is included in gas mixtures in some applications, especially in the case of red meat, where the color can only be preserved by oxygen, or fish and seafood, to avoid the growth of anaerobic pathogens such as *Clostridium*. Furthermore, it is essential for plant respiration.

Other gases

Helium is a light gas, absent in air, which is added to gas mixtures to detect leaks.

Argon has useful physical properties: completely inert, it is denser and more soluble than nitrogen, which allows for more effective purging and lower gas consumption. It also inhibits plant respiration.

Nitrous oxide is used in the expansion of foams and creams. It also inhibits the secretion of ethylene by plants, which has a cicatrizing effect on fruit skin. It is also bacteriostatic and fungistatic, which is useful for plant preservation. However, it is a greenhouse gas and has a negative image among consumers.

	N_2	CO_2	O_2	He	Ar	N_2O
Mechanical strength	☑					
Color preservation			☑			
Antioxidant	☑				☑	
Solubility		☑			☑	☑
Leak detection				☑		
Anti-aerobic activity		☑				☑
Anti-anaerobic activity		☑	☑			
Reduction in respiration		☑			☑	☑
Delayed ripening						☑
Oxygenation			☑			
Acidification		☑				
Expansion						☑

Table 8.8. *Main properties and applications of gases used in the food industry*

8.5.2.2. *Applications*

The choice of modified atmosphere for a given product is closely linked to its a_w, which determines its sensitivity to various spoilage agents (see Volume 1 [JEA 16a]). Modified atmosphere packaging applications can therefore be divided into three food categories: low moisture foods, intermediate moisture foods and high moisture foods [DRU 98].

Low moisture/dry foods (a_w < 0.4 – 0.5)

The shelf life of these foods (milk powder, potato flakes, dehydrated vegetables, coffee, oilseeds, dried fruit, etc.) is generally limited by oxidation and fat rancidity: nitrogen, used to reduce the oxygen content in the package, is therefore commonly used. Argon can also be used to deoxygenate the atmosphere in the package. The shelf life of such products stored at room temperature can therefore be tripled or quadrupled without using antioxidants.

Intermediate moisture foods (0.4 < a_w < 0.8)

The preservation of these products is more difficult because, in addition to inhibiting oxidation, it is also necessary to prevent enzymatic spoilage and mould growth. Pure CO_2 or N_2/CO_2 mixtures with a high percentage of CO_2 (needed against moulds) are often used. The percentage of each gas depends on many other factors such as the type of contaminating flora, the initial microbial load of the product, hygiene and handling, additive content, storage temperature, etc.

This category includes dough products, meat pastries, industrial pastries, etc. An oxygen absorber can be added to the gas mixture in the case of breads/pastries to prevent any unwanted changes in the package atmosphere.

High moisture foods (a_w > 0.8)

The risk of microbial growth is high in this category. CO_2 is always necessary, but the composition of the mixture depends on the product and user requirements (packaging type, shelf life, etc.). Many factors influence the effectiveness of the packaging, with the most important being temperature control.

This category includes most food products:

– *fresh charcuterie (prepared meat products)*: N_2/CO_2 binary gas mixtures extend the BBD (best before dates) of these products (fresh or cooked sausages, ham, blood pudding, etc.). Compared to vacuum packaging, modified atmosphere packaging has a more attractive presentation (fresh product image, products not stuck together, better color preservation).

– *poultry*: 50% N_2/50% CO_2 gas mixtures used to package whole poultry or poultry pieces double their shelf life.

– *cut of meat* (Table 8.9): the bright red color is a key factor influencing the consumer choice of beef. Packaging must therefore preserve this color while preventing microbial growth. To do this, it is necessary to use 60 to 70% O_2 to maintain myoglobin in an oxidized state, and 20 (minimum effectiveness) to 25% CO_2 (precipitation threshold of water soluble proteins responsible for browning). A mixture of 70% O_2/20% CO_2/10% N_2 preserves color and microbiological quality for a period of 21 days.

Product	*Shelf life in air*	*Shelf life in gas mixture*
Beef	4 days	10–15 days
Pork	4 days	6–9 days
Minced steak	2 days	4 days
Giblets (liver)	1–2 days	6 days

Table 8.9. *Comparison of shelf lives of different meats packaged in air and modified atmosphere (66% O_2, 25% CO_2, 9% N_2) and stored at 4°C*

– *fish*: smoked and salted fish are packaged in a mixture of N_2/CO_2 (with 30 to 50% CO_2), but the packaging of fresh fish poses a particular problem. The microbial contamination of the product is relatively high due to gutting and filleting, and enzymatic alteration is facilitated by the growth of aerobic microorganisms present in the mucus or spores of strictly anaerobic microorganisms. Fish is therefore packaged under good hygiene conditions in CO_2 modified atmosphere (50% N_2/50% CO_2) where the temperature is maintained close to 0°C and the product is very fresh (within three days of capture). It is then possible to achieve shelf lives of about six to eight days. If there is any doubt about the quality of the product, it is recommended to use a ternary mixture with 5 to 10% O_2 to avoid the growth of *Clostridium*.

For all these products, the most impermeable materials are used in modified atmosphere packaging in order to preserve the gas mixture proportions. This is not the case with fresh-cut vegetables, which undergo respiratory exchanges with their environment (see Volume 3 [JEA 16b]); these exchanges should be kept at a low rate so as not to suffocate the product (fermentative metabolism in the absence of oxygen), while ensuring compatibility between the packaging material, the product and the gas mixture. Plant metabolism causes a change in the composition of the atmosphere, which depends on plant respiration and the diffusive properties of the film. Creating an optimum atmosphere requires suitable films with a wide range of permeability to gases (oxygen, carbon dioxide and in some cases ethylene) and water vapor to cover all the respiration mechanisms of different plants [RIQ 94, VAR 00].

8.5.2.3. *Legislation*

EU Directive 94/54/EC requires that labeling must include the phrase "packaged in a protective atmosphere". Directive 95/2/EC, as amended by Directives 96/85/EC, 98/72/EC and 2001/8/EC, defines packaging gases as additives and only permits carbon dioxide, nitrogen, oxygen, argon, helium and nitrous oxide as packaging gases, propellants or acidity regulators, without restriction or maximum level. The purity criteria for these six gases are defined by Directive 96/77/EC, as amended by Directives 96/86/EC and 2000/63/EC. In addition, Directive 2001/5/EC allows hydrogen as a food additive. None of these directives defines standards for the microbiological quality of packaging gases.

Bibliography

[APL 98] APLINCOURT M., MARSAL P., PRUDHOMME J.C., "Interactions physicochimiques entre matériaux d'emballage métallique et constituants alimentaires : corrosion et protection" in MULTON J.L., BUREAU G (eds), *L'emballage des denrées alimentaires de grande consommation*, Lavoisier, Paris, pp. 263–295, 1998.

[BIM 95] BIMBENET J.J., LONCIN M., *Bases du Génie des Procédés Alimentaires*. Masson, Paris, 1995.

[BIM 02] BIMBENET J.J., DUQUENOY A., TRYSTRAM G., *Génie des Procédés Alimentaires: Des Bases aux Applications*, Dunod, Paris, 2002.

[BLA 82] BLANKENSHIP L.C., CRAVEN S.E., "Survival of *Campylobacter jejuni* in chicken meats as a function of temperature", *Appl. Microbiol*, vol. 44, pp. 88–92, 1982.

[BOU 96] BOURGEOIS C.M., LARPENT J.P., *Microbiologie Alimentaire, Tome 2: Aliments Fermentés et Fermentations Alimentaires*, Lavoisier, Paris, 1996.

[BRA 77] BRADSHAW J.G., PEELER J.T., TWEDT R.M., "Thermal inactivation of ileal loop-reactive *Clostridium perfringens* type A strains in phosphate buffer and beef gravy", *Appl. Environ. Microbiol*, vol. 34, pp. 280–284, 1977.

[BRA 85] BRADSHAW J.G., PEELER J.T., COORWING J.J. *et al.*, "Thermal resistance of *Listeria monocytogenes* in milk", *J. Food Protect*, vol. 48, pp. 743–745, 1985.

[BRO 99] BROCHETTE P., "Emulsification – élaboration et étude des emulsions", *Techniques de l'Ingénieur*, J2150, 1999.

[BRU 06] BRUMMER R., *Rheology Essentials of Cosmetic and Food Emulsions*, Springer Science & Business Media, 2006.

[CAN 79] CANTE C.J., FRANZEN R.W., SALEEB F.Z., "Proteins as emulsifiers: methods for assessing the role", *J. Am. Oil Chem. Soc.*, vol. 56, pp. 71–77, 1979.

[CAS 07] CASTLE L., "Chemical migration into food: an overview", in BARNES K., SINCLAIR R., WATSON D. (eds), *Chemical Migration and Food Contact Materials*, Woodhead Publishing, 2007.

[CHE 76] CHEFTEL J.C., CHEFTEL H., *Introduction à la biochimie et à la technologie des aliments*, Lavoisier, Paris, vol. 1, 1976.

[CHE 85] CHEFTEL J.C., CUQ J.L., LORIENT D., *Protéines alimentaires: Biochimie, propriétés fonctionnelles, valeurs nutritionnelles, modifications chimiques*, Lavoisier, Paris, 1985.

[CHE 81] CHERRY J.P., MC WATTERS K.H., "Whippability and aeration", in CHERRY J.P. (ed.), *Protein Functionality in Foods*, Am. Chem. Soc., Washington DC, 1981.

[CTC 97] CTCPA, Barêmes de Stérilisation pour Aliments Appertisés, Centre Technique de la Conservation des Produits Agricoles, Paris, 1997.

[COU 98] COUTURIER Y., "Les matériaux cellulosiques", in MULTON J.L., BUREAU G. (eds), *L'emballage des denrées alimentaires de grande consommation*, Lavoisier, Paris, 1998.

[DAI 08] DAINELLI D., GONTARD N., SPYROPOULOS D. *et al.*, "Active and intelligent food packaging: legal aspects and safety concerns", *Trends in Food Science & Technology*, vol. 19, pp. 103–112, 2008.

[DEL 15] DELAPLACE G., LOUBIERE K., DUCEPT F. *et al.*, *Dimensional analysis of food processes*, ISTE Press, London and Elsevier, Oxford, 2015.

[DEL 98] DE LEIRIS J.P. "Les films plastiques", in MULTON J.L., BUREAU G. (eds), *L'emballage des denrées alimentaires de grande consommation*, Lavoisier, Paris, 1998.

[DER 09] DE REYNAL B, MESCLE J.F. "Additifs conservateurs (antibactériens, antifongiques)", in DE REYNAL B., MULTON J.L. (eds), *Additifs et auxiliaires de fabrication dans les industries agroalimentaires*, Lavoisier, Paris, 2009.

[DER 09] DE REYNAL B., MULTON J.L., *Additifs et auxiliaires de fabrication dans les industries agroalimentaires*, 4th edition, Lavoisier, Paris, 2009.

[DEN 71] DENNY C.B. "Effect of toxin concentration on the inactivation of *Staphylococcal enterotoxin* A in beef bouillion and in phosphate buffer", *Appl. Microbiol.*, vol. 21, pp. 1064–1066, 1971.

[DIC 92] DICKINSON E., *An Introduction to Food Colloids*, Oxford University Press, 1992.

[DRU 98] DRULHE E., "Emballage sous atmosphère modifiée", in MULTON J.L., BUREAU G. (eds), *L'emballage des denrées alimentaires de grande consommation*, Lavoisier, Paris, 1998.

[EMB 12a] EMBLEM A., "Packaging functions", in EMBLEM A., EMBLEM H. (eds), *Packaging Technology: Fundamentals, Materials and Processes*, Woodhead Publishing, 2012.

[EMB 12b] EMBLEM H., "Packaging and environmental sustainability", in EMBLEM A., EMBLEM H. (eds), *Packaging Technology: Fundamentals, Materials and Processes*, Woodhead Publishing, 2012.

[FEI 98] FEIGENBAUM A., "Evaluation de la migration des matériaux plastiques au contact des aliments par des méthodes alternatives", in MULTON J.L., BUREAU G. (eds), *L'emballage des denrées alimentaires de grande consommation*, Lavoisier, Paris, 1998.

[GEN 95] GENIN N., RENE F., "Analyse du rôle de la transition vitreuse dans les procédés de conservation agroalimentaires", *J. Food Eng.*, vol. 26, pp. 391–408, 1995.

[GON 94] GONTARD N., GUILBERT S. "Bio-packaging : technology and properties of edible and/or biodegradable material of agricultural origin", in MATHLOUTHI M. (ed.), *Food Packaging and Preservation*, Springer, 1994.

[GRA 12] GRAYHURST P., "Glass packaging", in EMBLEM A., EMBLEM H. (eds), *Packaging Technology: Fundamentals, Materials and Processes*, Woodhead Publishing, 2012.

[GUE 98] GUERIN J., "Emballage pour cuisson sous vide", in MULTON J.L., BUREAU G. (eds), *L'emballage des denrées alimentaires de grande consommation*, Lavoisier, Paris, 1998.

[HAL 81] HALLING P.J., "Protein-stabilized foams and emulsions", *C.R.C. Crit. Rev. Food Sci. Nutr.*, vol. 15, pp. 155–203, 1981.

[HAL 88] HALLSTRÖM B., SKJÔLDEBRAND C., TRÄGARDH C., *Heat Transfer and Food Products*, Elsevier, London, 1988.

[HAR 92] HARTEL R.W., "Evaporation and freeze concentration", in HELDMAN D.R., LUND D.B. (eds), *Handbook of Food Engineering*, M. Dekker, New York, USA, 1992.

[JAS 94] JASSE B., SEUVRE A.M., MATHLOUTHI M., "Permeability and structure in polymeric packaging materials", in MATHLOUTHI M. (ed.), *Food Packaging and Preservation*, Springer, 1994.

[JEA 11] JEANTET R., BRULE G., DELAPLACE G., *Génie des procédés appliqué à l'industrie laitière*, Tec & Doc éditions, Paris, 2011.

[JEA 16a] JEANTET R., CROGUENNEC T., SCHUCK P, BRULÉ G. (eds), *Handbook of Food Science and Technology 1: Food Alteration and Food Quality*, ISTE, London and John Wiley and Sons, New York, 2016.

[JEA 16b] JEANTET R., CROGUENNEC T., SCHUCK P, BRULÉ G. (eds), *Handbook of Food Science and Technology 3: Food Biochemistry and Technology*, ISTE, London, and John Wiley & Sons, New York, 2016.

[KAR 95] KARBSTEIN H., SCHUBERT H., "Developments in the continuous mechanical production of oil-in-water macro-emulsions", *Chem. Eng. Proc.*, vol. 34, pp. 205–211, 1995.

[KES 86] KESSLER H.G., "Energy aspects of food preconcentration", in MAC CARTHY D. (ed.), *Concentration and Drying of Foods*, Elsevier, London, England, 1986.

[KIN 81] KINSELLA J.E., "Functional properties of proteins: possible relationships between structure and function in foams", *Food Chem.*, vol. 7, pp. 273–288, 1981.

[KNO 90] KNOCKAËRT C., *Le Fumage du Poisson*, 2nd edition, IFREMER, 1990.

[KOI 83] KOIDIS, P., DOYLE M.P., "Survival of *Campylobacter jejuni* in fresh and heated red meat", *J. Food Protect.*, vol. 46, pp. 771–774, 1983.

[KRO 77] KROG N., "Functions of food emulsifiers in food systems", *J American Oil Chem. Soc.*, vol. 54, pp. 124–131, 1977.

[KRO 76] KROG N., LAURIDSEN J., "Food emulsifiers and their associations with water", in FRIBERG S. (ed.), *Food Emulsions*, Marcel Dekker Inc, New York, 1976.

[LAC 92] LACROIX J.P., SINAI C., "Preservation of food by ionization", *Ind Alim. Agric.*, vol. 109, no. 10, pp. 701–703, 1992.

[LEL 92] LELEU R., "Opérations unitaires: Evaporation", *Techniques de l'ingénieur*, J 2320, 1992.

[LEB 98] LE BARS M., BOUROCHE A., *L'ionization dans l'Industrie Agroalimentaire*, INRA editions, 1998.

[LON 61] LONCIN M., *Les Opérations Unitaires du Génie Chimique*, Dunod, Paris, 1961.

[LON 76] LONCIN M., *Génie Industriel Alimentaire, Aspects Fondamentaux*, Masson, Paris, 1976.

[LOV 82] LOVETT J., BRADSHAW J.G., PEELER J.T., "Thermal inactivation of *Yersinia enterocolitica* in milk", *Appl. Environ. Microbiol.*, vol. 44, pp. 517–519, 1982.

[LOX 98] LOX F., PASCAT B., "Qualité des emballages : migration", in MULTON J.L., BUREAU G. (eds), *L'emballage des denrées alimentaires de grande consommation*, Lavoisier, Paris, 1998.

[MAF 96] MAFART P., "Les procédés physiques de conservation", *Génie Industriel Alimentaire*, Tec et Doc éditions, Paris, p. 295, 1996.

[MAL 06] MALKIN A.Y., ISAYEV A.I., *Rheology: Concepts, Methods & Applications*, ChemTec Publishing, 2006.

[MCC 99] MCCLEMENTS D.J., *Food Emulsions: Principles, Practice, and Techniques*, Boca Raton, Florida, 1999.

[MUL 98] MULTON J.L., BORGHESE T., SAUVAGE F. *et al.*, "Les fonctions de l'emballage", in MULTON J.L., BUREAU G. (ed.), *L'emballage des denrées alimentaires de grande consommation*, Lavoisier, Paris, 1998.

[PAG 12] PAGE B. "Rigid metal packaging", in EMBLEM A., EMBLEM H. (eds), *Packaging Technology : Fundamentals, Materials and Processes*, Woodhead Publishing, 2012.

[RAF 98] RAFFI J. "Identifying irradiated foods", *Trends Anal. Chem.*, vol. 17, no. 4, pp. 226–233, 1998.

[REA 66] READ R.B., BRADSHAW J.G. "Staphylococcal enterotoxin B thermal inactivation in milk", *J. Dairy Science*, vol. 49, no. 2, pp. 202–203, 1966.

[RIL 12a] RILEY A. "Plastics manufacturing processes for packaging materials", in EMBLEM A., EMBLEM H. (eds), *Packaging Technology: Fundamentals, Materials and Processes*, Woodhead Publishing, 2012.

[RIL 12b] RILEY A. "Paper and paperboard packaging", in EMBLEM A., EMBLEM H. (eds), *Packaging Technology: Fundamentals, Materials and Processes*, Woodhead Publishing, 2012.

[RIQ 94] RIQUELME F., PRETEL M.T., MARTINEZ G. *et al.*, "Packaging of fruits and vegetables : recent results", in MATHLOUTHI M. (ed.), *Food Packaging and Preservation*, Springer, 1994.

[ROB 93] ROBIN O., REMILLARD N., PAQUIN P. "Influence of major process and formulation parameters on microfluidized fat globule size distribution and exemple of a practical consequence", *Colloids and Surfaces A: Physicochemical Engineering Aspects*, vol. 80, pp. 211–222, 1993.

[SCH 07] SCHÄFER A., "Regulation of food contact materials in the EU", in BARNES K., SINCLAIR R., WATSON D. (eds), *Chemical Migration and Food Contact Materials*, Woodhead Publishing, 2007.

[SCH 97] SCHUBERT H., STANG M., "New developments in the production of food emulsions", *2nd Congrès Mondial de l'Emulsion*, Bordeaux, 1997.

[SHE 76] SHEPHERD I., YOELL W., "Cake emulsions", in FRIBERG S. (ed.), *Food Emulsions*, Marcel Dekker Inc, New York, 1976.

[SIV 12] SIVERTSVIK M., ROSNES J.T., BERGSLIEN H., "Modified atmosphere packaging", in OHLSSON T., BENGTSSON N. (eds), *Minimal Processing Technologies in the Food Industries*, Woodhead Publishing, 2012.

[SWE 83] SWEETSER M.A., MUIR D.D., "Effect of homogenization on the heat stability of milk", *J. Dairy Res.*, vol. 50, pp. 291–300, 1983.

[VAR 00] VAROQUAUX P., "Les films à perméabilité aux gaz ajustable : application aux fruits et légumes", in GONTARD N. (ed.), *Les emballages actifs*, Lavoisier, Paris, pp. 89–108, 2000.

[VAS 84] VASSEUR J., BIMBENET J.J., DUPRAT J.C. *et al.*, "Séchage sur cylindres dans les IAA : nouvelles techniques, nouvelles perspectives", *Ind Alim Agric*, vol. 101, pp. 911–919, 1984.

[WAG 89] WAGNER J., GILL, C.O., "Packaging of meat for prolonged chilled storage", *British Food J.*, vol. 91, pp. 11–15, 1989.

[WAL 75] WALSTRA P., "Effect of homogenization on the fat globule size distribution in milk", *Neth. Milk Dairy J.*, vol. 29, pp. 279–294, 1975.

[WAL 93] WALSTRA P., "Principles of emulsion formation", *Chem. Eng. Sci.*, vol. 48, pp. 333–349, 1993.

[WAL 98] WALSTRA P., SMULDERS P.E.A., "Emulsion formation", in BINKS B. (ed.), *Modern Aspects of Emulsion Science*, Royal Society of Chemistry, Cambridge, UK, 1998.

[WES 94] WESTERGAARD V., "*Milk Powder Technology, Evaporation and Spray Drying*, 4th edition, Niro A/S, Copenhagen, Denmark, 1994.

[WOO 79] WOODBURN M.J., SOMERS E., RODRIGUEZ J. *et al.*, "Heat inactivation rates of botulism toxins A, B, E and F in some foods and buffers", *J. Food Sci.*, vol. 44, pp. 1658–1661, 1979.

List of Authors

Gérard BRULÉ
Agrocampus Ouest
Rennes
France

Thomas CROGUENNEC
Agrocampus Ouest
Rennes
France

Juliane FLOURY
Agrocampus Ouest
Rennes
France

Romain JEANTET
Agrocampus Ouest
Rennes
France

Valérie LECHEVALIER
Agrocampus Ouest
Rennes
France

Pierre SCHUCK
Institut national de la recherche agronomique (INRA)
Rennes
France

Index

B, C

D

E, F

G, H

I, L, M

N, O, P

R, S, T

U, V, W

Printed and bound by CPI Group (UK) Ltd, Croydon, CR0 4YY
13/09/2026
14961025-0003